THE ENCYCLOPEDIA OF
BIRDS

THE ENCYCLOPEDIA OF
BIRDS

Volume 4

An imprint of Infobase Publishing

THE ENCYCLOPEDIA OF BIRDS

Copyright ©2007 by International Masters Publishing.

Portions of this material were previously published
as part of the *Wildlife Explorer* reference set.

All rights reserved. No part of this book may be reproduced or utilized in any form
or by any means, electronic or mechanical, including photocopying, recording, or
by any information storage or retrieval systems, without permission in writing
from the publisher. For information contact:

Facts On File, Inc.
An imprint of Infobase Publishing
132 West 31st Street
New York NY 10001

ISBN-10: 0-8160-5904-7 (set)
ISBN-13: 978-0-8160-5904-1 (set)

Library of Congress Cataloging-in-Publication Data
The encyclopedia of birds / edited by International Masters Publishers.
p. cm.
Includes index.
ISBN 0-8160-5904-7 (set)
1. Birds—Encyclopedias. I. International Masters Publishers.
QL672.2E534 2006
598.03—dc22   2006049526

Facts On File books are available at special discounts when purchased in bulk quantities
for businesses, associations, institutions, or sales promotions. Please call our
Special Sales Department in New York at (212) 967-8800 or (800) 322-8755.

You can find Facts On File on the World Wide Web at http://www.factsonfile.com

Editorial Director: Laurie E. Likoff
Project Editor: Tracy Bradbury
Text and cover design by Cathy Rincon

Printed in China

CP FOF 10 9 8 7 6 5 4 3 2 1

This book is printed on acid-free paper.

---

*The Publisher has made every effort to contact and secure permission releases
from the copyright holders of the photographs used throughout this work.
Anyone having claims to ownership not identified in the picture credits is invited
to email to photos@impdirect.com.*

CONTENTS

Introduction
vii

What is a Bird?
viii

List of Birds
ix

House Sparrow to Peregrine Falcon
533–709

INTRODUCTION

The Encyclopedia of Birds is a six-volume set designed to introduce the young reader to the fascinating world of birds. Birds, in all their variety, from the forests of North America, to the beaches of South America, to the mountains of Europe and the plains of Australia, share certain common features of anatomy and physiology as well as habitat and breeding. But there are also significant differences among the populations as well as unique relationships in courtship routines, nesting and life expectancy.

In this reference work, the birds are arranged alphabetically and appear in four-page spreads.

Each bird featured includes the following information:

- Order, family, genus and species
- Habitat
- Behavior
- Breeding
- Food and Feeding
- Distribution and Range
- Vital Statistics
- Creature Comparisons

In addition, fun facts or unusual information is imparted in the "Did You Know?" box and sidebar information often includes notes on conservation, related species, unusual behavior or distinguishing features.

An information panel in each section includes vital statistics on weight, length, wingspan, sexual maturity, breeding season, number of eggs, incubation period, fledging period, typical diet and lifespan.

Richly enhanced by full-color photographs as well as drawings and labeled diagrams, this wide-ranging set will be sure to fascinate and entertain bird lovers of all ages.

—Laurie E. Likoff
Editorial Director

WHAT IS A BIRD?

Generally speaking a bird is any member of the class known as Aves that share certain common characteristics and traits. Birds are warm-blooded, bipedal animals whose anatomy is characterized by forelimbs modified through natural selection and evolution to become wings, whose exterior is covered by feathers, and that have, in most cases, hollow bones to assist in flight.

Most birds are diurnal, or active during the day, but some are nocturnal, active during the evening hours, such as owls, and still others feed either day or night as needed.

Many birds migrate long distances to find the optimum or ideal habitats, while others rarely range from their original breeding spots.

Shared characteristics of birds may include a bony or hard beak with no teeth, the laying of hard-shelled eggs, a light but strong skeleton, and a high rate of metabolism.

Most birds are characterized by flight although several well-known species, particularly those that reside on islands, have now lost this ability. Some common flightless birds include the ostrich, penguin, kiwi and now extinct Dodo.

Birds feed on plants, seeds, insects, fish, carrion or other birds. Birds are also an important food source for humans. The most commonly eaten species is the domestic chicken, although geese, pheasants, turkeys and ducks are also common fare, particularly around Thanksgiving Day and the holidays. Birds grown for human consumption are known as poultry. Humans have caused the disappearance of some species due to habitat destruction, hunting or over consumption.

Other species of birds have come to depend on human activities for food and are so widespread as to be considered a nuisance such as the common pigeon or rock pigeon. In North America, sparrows, starlings, and finches are also widespread. Some birds have been used by humans to perform tasks, such as homing pigeons in the days before modern communications, and falcons to aid in hunting or for sport. Tropical birds are often sought after and kept as pets although some are now listed as endangered and their trafficking for this purpose has been restricted.

The bird population, like many other fish and wildlife groups, is facing threats worldwide. According to Worldwatch Institute, bird populations are declining, with 1,200 species facing extinction in the next century. Among the most prevalent reasons cited are habitat loss, predation by nonnative species, oil spills and pesticide use, climate change and excessive rates of hunting and fishing. All these threats make it ever more important to understand, appreciate and protect the birds we see around us everyday.

—Kenny Clements

LIST OF BIRDS

Adelie penguin	1	Blue-and-yellow macaw	141	Dovekie	277
African fish eagle	5	Blue-footed booby	145	Dwarf cassowary	281
African gray parrot	9	Blue tit	149	Eagle owl	285
African harrier hawk	13	Boat-billed heron	153	Eastern screech owl	289
African pygmy falcon	17	Bohemian waxwing	157	Eclectus parrot	293
American black vulture	21	Broad-billed		Egyptian plover	297
American cliff swallow	25	hummingbird	161	Egyptian vulture	301
American coot	29	Brown creeper	165	Elf owl	305
American harpy eagle	33	Brown pelican	169	Emperor penguin	309
American redstart	37	Brown skua	173	Emu	313
American robin	41	Budgerigar	177	Eurasian avocet	317
American swallow-tailed		Burrowing owl	181	Eurasian buzzard	321
kite	45	Canada goose	185	Eurasian cuckoo	325
Andean Condor	49	Carrion crow	189	Eurasian curlew	329
Anhinga	53	Cattle egret	193	Eurasian kingfisher	333
Arctic tern	57	Chaffinch	197	Eurasian nuthatch	337
Atlantic puffin	61	Chiffchaff	201	Eurasian oystercatcher	341
Bald eagle	65	Cockatiel	205	Eurasian robin	345
Bananaquit	69	Common eider	209	Eurasian sparrowhawk	349
Bank swallow	73	Common grackle	213	Eurasian swift	353
Barn owl	77	Common guillemot	217	Eurasian woodcock	357
Barn swallow	81	Common kestrel	221	European starling	361
Bateleur	85	Common loon	225	Feral pigeon	365
Bee hummingbird	89	Common nighthawk	229	Flightless cormorant	369
Belted kingfisher	93	Common peafowl	233	Galapagos ground finches	373
Black-billed magpie	97	Common pheasant	237	Goldcrest	377
Black-browed albatross	101	Common quail	241	Golden eagle	381
Black-capped chickadee	105	Common snipe	245	Golden oriole	385
Black-crowned night heron	109	Common tailorbird	249	Golden pheasant	389
Black-headed gull	113	Common yellowthroat	253	Goliath heron	393
Black-winged stilt	117	Cooper's hawk	257	Gouldian finch	397
Black heron	121	Corncrake	261	Gray-crowned crane	401
Black skimmer	125	Crab plover	265	Gray heron	405
Black tern	129	Dipper	269	Gray partridge	409
Black woodpecker	133	Double-wattled		Graylag goose	413
Blackbird	137	cassowary	273	Great argus pheasant	417

Great black-backed gull	421	Marabou stork	613	Ruddy duck	801
Great bustard	425	Martial eagle	617	Ruddy turnstone	805
Great cormorant	429	Mute swan	621	Ruff	809
Great crested grebe	433	Namaqua sandgrouse	625	Rufous hummingbird	813
Great curassow	437	New Holland honeyeater	629	Sacred ibis	817
Great gray owl	441	Nightingale	633	Sacred kingfisher	821
Great gray shrike	445	North American bittern	637	Saddle-bill stork	825
Great horned owl	449	Northern cardinal	641	Satin bowerbird	829
Great Indian hornbill	453	Northern flicker	645	Scarlet ibis	833
Great spotted woodpecker	457	Northern gannet	649	Screamers	837
Great white pelican	461	Northern goshawk	653	Secretary bird	841
Greater flamingo	465	Northern harrier	657	Shoebill	845
Greater prairie chicken	469	Northern mockingbird	661	Short-toed snake eagle	849
Greater rhea	473	Northern pintail	665	Skylark	853
Greater roadrunner	477	Northern wheatear	669	Slavonian grebe	857
Green jay	481	Oilbird	673	Smew	861
Guianan cock-of-the-rock	485	Osprey	677	Snail kite	865
Gyrfalcon	489	Ostrich	681	Snow bunting	869
Hamerkop	493	Ovenbird	685	Snow goose	873
Harlequin duck	497	Oxpeckers	689	Snowy owl	877
Hawfinch	501	Painted bunting	693	Snowy sheathbill	881
Helmeted guineafowl	505	Palm cockatoo	697	Sooty tern	885
Herring gull	509	Paradise whydah	701	Southern giant petrel	889
Hill mynah	513	Pel's fishing owl	705	Southern ground	
Himalayan snowcock	517	Peregrine falcon	709	hornbill	893
Hoatzin	521	Pheasant coucal	713	Southern yellow-billed	
Hooded vulture	525	Pheasant-tailed jacana	717	hornbill	897
Hoopoe	529	Raggiana bird of paradise	721	Spotted nutcracker	901
House sparrow	533	Rainbow bee-eater	725	Stone curlew	905
Hyacinth macaw	537	Rainbow lorikeet	729	Sulphur-crested cockatoo	909
Japanese crane	541	Raven	733	Sun bittern	913
Kakapo	545	Razorbill	737	Superb lyrebird	917
Kea	549	Red crossbill	741	Swallow-tailed gull	921
King penguin	553	Red Junglefowl	745	Tawny frogmouth	925
King vulture	557	Red kite	749	Tinamou	929
Kiwis	561	Red-and-yellow barbet	753	Toco toucan	933
Kori bustard	565	Red-billed quelea	757	Torrent duck	937
Lammergeier	569	Red-breasted goose	761	Trumpeter swan	941
Lappet-faced vulture	573	Red-breasted merganser	765	Tufted duck	945
Laughing kookaburra	577	Red-headed woodpecker	769	Tundra swan	949
Lovebirds	581	Red-tailed tropicbird	773	Turtle dove	953
Luzon bleeding-heart	585	Red-winged blackbird	777	Verreaux's eagle	957
Macaroni penguin	589	Reddish egret	781	Victoria crowned pigeon	961
Magnificent frigatebird	593	Reed warbler	785	Village weaver	965
Mallard	597	Rockhopper penguin	789	Wallcreeper	969
Malleefowl	601	Rose-ringed parakeet	793	Wandering albatross	973
Mandarin duck	605	Ruby-throated		Water rail	977
Manx shearwater	609	hummingbird	797	Waved albatross	981

x List of Birds

Weka	985	White stork	1009	Wood pigeon	1033
Western capercaillie	989	Wild turkey	1013	Wood stork	1037
Western tanager	993	Willow ptarmigan	1017	Wryneck	1041
Whimbrel	997	Winter wren	1021	Yellow-billed cuckoo	1045
Whippoorwill	1001	Wompoo fruit dove	1025		
White-fronted bee-eater	1005	Wood duck	1029		

House Sparrow

- **ORDER** -
Passeriformes

- **FAMILY** -
Passeridae

- **GENUS & SPECIES** -
Passer domesticus

KEY FEATURES

● One of the world's most widespread and successful songbirds

● Able to live in every habitat within its range, except for dry deserts, thick forests and mountain peaks

● Enjoys a closer relationship with humans than does almost any other species of bird, and has actually benefited from the growth of cities and towns

WHERE IN THE WORLD?

Formerly restricted to North Africa and Eurasia, but has spread throughout the world during the last 150 years; now represented on six continents; absent only from polar extremes

House Sparrow 533

LIFECYCLE

BEHAVIOR

Most animals are made homeless when their habitats are replaced by farmland and towns — but the house sparrow has benefited greatly from cultivation and urbanization.

HABITAT

Few parts of the world are too hot, cold, wet or dry for the house sparrow. It is found well above the Arctic Circle in Scandinavia, and at the equator in South America and eastern Africa.

But the species was not always so widespread. Until the early 19th century, the sparrow was confined to lowlands in southern and central Eurasia and to a few parts of North Africa. Since then, urbanization and the planting of cereal crops, like wheat, have enabled this grain eater to colonize much of the world.

▼ **WORLD RESIDENT**
The sparrow can survive almost anywhere.

It is unusual to see a house sparrow on its own; the species feeds in parties, breeds in loose colonies and roosts communally. One reason for flocking together is to increase the chance of spotting danger. Additionally, since each sparrow in the flock spends less time checking for predators, it can feed longer. The house sparrow has many enemies, even in towns; a large number of nestlings and juveniles falls victim to cats.

Breeding sparrows sleep near the nest. At other times, they assemble in huge numbers at favored sites: a roost in Egypt housed about 100,000 birds. By repeating a chirping call, a flock of roosting sparrows is capable of producing a deafening noise.

The house sparrow frequents many dusty habitats, so it must clean its plumage regularly. Flocks often settle to bathe in park lakes, fountains and puddles. The sparrow also takes dust baths, flicking sand and soil with its wings, to remove parasites. Occasionally it perches on the rim of a chimney to expose its feathers to the smoke.

▶ **TOP SEED**
The house sparrow is a pest in some regions due to its eating habits.

 *DID YOU KNOW?*

● The house sparrow has learned that insects trapped in spiders' webs and car radiators make an easy meal.

● The house sparrow annoys gardeners by tearing and shredding yellow flower petals, for which it has a fondness.

BREEDING

FOOD & FEEDING

The house sparrow feeds mainly on grass seeds, herbs, wildflowers and cereal crops, especially wheat, barley, oats, millet and sorghum. It gathers food by pecking on the ground, but also tugs the seeds from low-growing plants and perches on ripe seed heads to strip them. Large flocks gather to feast in ripening corn fields, becoming a pest and occasionally destroying whole crops. Flocks also raid grain stored at farmyards, railway depots and ports, even stealing food from cattle in winter.

The house sparrow is an opportunist, quick to exploit any source of food. Depending on the season, it nibbles at fresh plant shoots and buds, extracts the sweet pulp from fruits and hunts insects.

In urban areas, food is plentiful all year. The house sparrow scavenges a wide range of scraps from shops, markets, parks and open-air cafés; it also visits bird feeders to take the food put out by humans. Indeed, some city-dwelling sparrows eat almost no "natural" food at all.

CONSERVATION

The house sparrow continues to expand its range alongside human expansion, especially in South America and oceanic islands. But in Britain, flock numbers have dropped from 19 birds in 1970 to 12 in 1995.

HOUSE PARTY

① Guests arrive…
A flock has discovered a garden bird feeder laden with seeds. The sparrows help themselves to this lucky find.

② Confrontation…
The sparrows squabble, jostling each other for the best positions. With his bill agape, one male threatens a rival.

③ Top feeder…
The aggressive sparrow's threat display and large bib keep the smaller-bibbed male away from the prime feeding spot.

④ Bonus
The top male gets the pick of the food and attracts the female, while the less showy male is left hungry and without a mate.

The house sparrow breeds in colonies of 10–20 birds. Nest sites vary from cliff faces to windowsills and rooftops. The sparrow may also take over the old nests of other birds; whole colonies can fit inside the bulky nests of herons or storks.

Both parents incubate the clutch of up to five eggs. To feed the chicks, the parents regurgitate a partly digested mixture of seeds and insects into the nestlings' bills. The young leave the nest after 2–3 weeks, when they learn to fly, but still beg food from their parents. After five weeks, the young become independent.

◄ **PAIR-BOND**
Both parents contribute to raising young.

▲ **ONE OF MANY**
In favorable conditions a female sparrow can raise seven broods in a year.

House Sparrow 535

PROFILE   HOUSE SPARROW

With a versatile, powerful bill and a broad diet, the resourceful house sparrow manages to thrive in a variety of habitats.

FEMALE
The female has a cream-colored stripe above each eye and a pale yellowish bill. She's usually smaller than the male.

COLORATION
Both sexes have patterned brown upperparts and pale underparts. The male (*right*) has a black bib, gray crown and rump, chestnut temples and a silver-gray bill.

BIB
Males have varying amounts of black on the throat and breast. Those with the largest black patches (bibs) have the highest levels of testosterone and are more attractive to females. Outside the breeding season, the bib is flecked with gray feathers.

BILL
The thick, cone-shaped bill can split seeds and tear morsels from scraps. The sparrow's muscular tongue helps it to maneuver seeds into position for cracking and swallowing.

FEET
The sparrow's relatively stout feet enable it to hop along the ground and grip perches securely, but they are of little use for walking or holding food.

CREATURE COMPARISONS

A bird of arid, sandy habitats, the golden sparrow (*Passer luteus*) is found in Africa from Mauritania and Senegal to southern Egypt, Sudan and Eritrea, with a smaller population in the Middle East.

The male has a yellow head, breast and belly, and his back is yellow (Arabian) or red-brown (African). The female is paler than the male, but is bright enough to make the female house sparrow seem drab by comparison. The golden sparrow is about 15% smaller than the house sparrow, with a shorter gray bill that turns black during the breeding season.

House sparrow

Golden sparrow

VITAL STATISTICS

WEIGHT	0.85–1.33 oz.
LENGTH	5.5–6"
WINGSPAN	8.5–10"
SEXUAL MATURITY	1 year
BREEDING SEASON	Spring and summer; all year in tropics
NUMBER OF EGGS	3–5 per clutch
INCUBATION PERIOD	9–18 days
FLEDGING PERIOD	11–19 days
BREEDING INTERVAL	2–4 broods a year; up to 7 a year in tropics
TYPICAL DIET	Seeds, berries, buds, insects; refuse scraps
LIFESPAN	1–5 years

RELATED SPECIES

● The sparrow family, *Passeridae*, contains 34 species in 3 genera: the house sparrow and 20 other species in the genus *Passer*; 6 species of rock and bush sparrow in the genus *Petronia* and 7 species of snow finch in the genus *Montifringilla*. Sparrows are related to weavers, queleas, bishops and fodies. As a family, they are generally brown or gray, with short legs and poor singing ability.

Hyacinth Macaw

• ORDER •	• FAMILY •	• GENUS & SPECIES •
Psittaciformes	Psittacidae	Anodorhynchus hyacinthinus

KEY FEATURES

- This rare, majestic blue bird is the longest of all parrots

- Does not reach breeding maturity until the age of 10 years

- Prefers to lay its eggs in the holes of dead palm tree stumps

- Tremendously strong beak is mainly used to crack open nuts — but is capable of tearing apart a wrought-iron cage

WHERE IN THE WORLD?

Found mainly in jungles and forests in central South America; through interior northeastern Brazil and extending into eastern Bolivia and northern Paraguay

Hyacinth Macaw 537

LIFECYCLE

Hyacinth macaw pairs form a close bond that can last for life; though they both care for their chicks through early life, the survival rate is low, due to predators and climate.

HABITAT

The hyacinth macaw is found mainly in the inland tropical jungles, swamps, forests and palm groves of Brazil. The three main areas of distribution include Amazonia in Brazil; around the pantanal habitats of the upper Rio Paraguay basin extending into Bolivia; and through interior northeastern Brazil.

▲ **CALM IN THE PALM**
A small flock of hyacinth macaws rests close to its abundant food source.

 DID YOU KNOW?

- The two-toned tongue of the hyacinth macaw is black and yellow.

- In addition to water, macaws drink fluid from unripe palm fruits.

- The hyacinth macaw rubs its two bill sections against each other; this not only sharpens them but also prevents excessive growth, which would impede feeding.

FOOD & FEEDING

The macaw eats a variety of fruits, nuts and seeds, but it favors the nuts of palm trees, using its strong bill to simultaneously crack a nut and mash the kernel. The fleshy tongue helps to keep large particles out of the throat. The macaw has discovered that rodents eat the outer layers from fallen nuts, but cannot manage to crack the kernels — the tasty portions the macaws prefer. The clever macaw cuts and drops fruits to the ground for the rodents, such as agoutis, then returns the following day to grab the dehusked kernels.

▶ **FAVORITE FOOD**
A macaw can turn in all directions to reach a nut.

BREEDING

In the early phases of courtship, the male hyacinth macaw will perform a number of displays for the female, such as dipping his head, spreading out his tail and lowering his wings; he will also contract his pupils and offer her food. The female will reciprocate with similar exhibitions. Both members of the pair will peck and nibble each other gently as signs of affection. The female nests high in a hollow tree hole just large enough for her to fit through; this helps protect her eggs from predators and temperature changes. She lays 2–3 white eggs, which she alone incubates for 27–30 days.

Young chicks are sensitive to the outside elements and do not leave the parents' care until about 12 weeks of age. It is rare for more than one youngster to be successfully reared to fledging because the chicks are especially susceptible to damp weather conditions and threats from poachers, who sell them for captive breeding. Juveniles reach maturity at 10 years of age.

 CONSERVATION

Considered rare and vulnerable, the hyacinth macaw has been on the IUCN (World Conservation Union) Red List in Bolivia, Brazil and Paraguay since 1994. With populations between 2,550–5,000, the hyacinth macaw is at risk of becoming extinct if humans continue to exploit its habitat. Illegal capture and hunting, as well as logging, agriculture and hydroelectric projects, are major threats to this bird's survival and habitats.

▶ **PECK AND PREEN**
A courting pair of hyacinth macaws affectionately nibble while cleaning hard-to-reach head feathers.

BEHAVIOR

Mostly sedentary, the hyacinth macaw circulates in small flocks or pairs into areas where food is available. Though capable of flying to great heights, the hyacinth macaw stays low, flapping its wings steadily. The small flocks will stay together during many night journeys, remaining active and quite visible on moonlit nights. In flight, the hyacinth macaw utters loud, repetitive cries and will draw from an extensive repertoire of sounds during courtship rituals, including *trarrree-arree* and *kru* calls. The sound of the bird's harsh warning calls has often been compared to the growling and whimpering of a small dog.

▲ **POWERFUL PERCHER**
The macaw grasps a branch with its thick toes.

BONDING TOGETHER

❶ Showing off…
Two macaws perch on a dead branch; one hangs upside down as the other watches with interest.

❷ House hunting…
The breeding pair searches for the perfect home. A tree cavity will provide adequate protection.

❸ Room for one more…
After laying and rearranging her eggs, the female remains in the snug nest for about 30 days.

❹ Caring parents
The adult birds steadfastly care for this one remaining chick, a survivor of damp weather and predators.

Hyacinth Macaw 539

PROFILE   HYACINTH MACAW

With its powerful bill and brilliant blue coloring, the majestic macaw is a portrait of strength and beauty.

SKULL
The powerful jaws are controlled by strong muscles used to crack open the toughest seeds.

EYES
Despite limited eye movement, the macaw's vision is good; the bird turns its head and neck to change its view. The black iris is circled by a yellow band.

PLUMAGE
The macaw's vivid violet-blue color stands out in its forest home. The undersides of its long, pointed wings and tail are black.

BILL & TONGUE
The stout, curved bill is hooked and hinged, which offers leverage for cracking nuts and tearing fruits. The prehensile tongue helps remove flesh from these foods.

PREENING GLAND
A preening gland, located at the base of the tail, supplies an oil that is applied to the feathers for grooming and insulation.

FEET
Two toes in the front and two in the back give the bird great agility in climbing and also may be used to grasp seeds and fruit while feeding.

CREATURE COMPARISONS

The great green macaw (*Ara ambigua*) is slightly smaller than its blue counterpart, the hyacinth macaw. The green macaw has much in common with its relative: it is very rare; it has a large head and hooked beak; and it favors nuts and seeds. The green macaw's plumage is mostly olive-green with a contrasting red and blue tail. Both species lay 2–3 eggs and prefer to travel in small family units or pairs. The great green macaw is found only in the Atlantic lowland of Central America in Costa Rica and in northern Colombia, generally farther north and west than the hyacinth macaw.

Great green macaw

Hyacinth macaw

VITAL STATISTICS

WEIGHT	3–4 lbs.
LENGTH	Up to 40", including a 24" tail
SEXUAL MATURITY	10 years
BREEDING SEASON	December–March
NUMBER OF EGGS	2–3
INCUBATION PERIOD	27–30 days
FLEDGING PERIOD	100–110 days
BREEDING INTERVAL	1 year
TYPICAL DIET	Nuts, seeds, fruit
LIFESPAN	Unknown

RELATED SPECIES

● The hyacinth macaw is 1 of 3 species in the genus *Anodorhynchus*; the others are the indigo macaw, *A. leari*, and the glaucous macaw, *A. glaucus*. All three species have striking blue plumage and reside in South America. There are 332 species in 78 genera of parrot in the *Psittacidae* family. Parrots join the cockatoos of the family *Cacatuidae*, in the order *Psittaciformes*.

JAPANESE CRANE

• ORDER •
Gruiformes

• FAMILY •
Gruidae

• GENUS & SPECIES •
Grus japonensis

KEY FEATURES

● Dances with its partner and other cranes, in rituals that extend beyond courtship into everyday life

● Graceful but raucous ballet seems sometimes to be performed out of sheer joy

● Japan's revered tancho, or "bird of happiness," this affectionate crane forms a lifelong pair bond and is upheld as a symbol of fidelity in marriage

WHERE IN THE WORLD?

Breeds as a resident species in the northeast of Hokkaido island, northern Japan; those that breed in eastern Siberia and northern China migrate to winter south in eastern China and Korea

LIFECYCLE

Almost all cranes "dance" to their partner, but few take their "fun" so seriously as the Japanese crane, which skips and frolics through almost every week of its life.

HABITAT

The Japanese crane lives in freshwater wetlands close to lakes and rivers. It favors marshy moorland and damp, sparsely wooded areas, where dense beds of tall reeds or grasses screen its nest.

About 600 cranes live on the Japanese island Hokkaido, one-third of the entire world population. They survive severe winters by moving to farmland, where they're artificially fed. Cranes in Siberia and China migrate south to spend winter on coastal marshes, mudflats and paddy fields.

For safe roosting, it prefers to stand in shallow water on submerged sandbars in rivers.

◀ **FIELD FARE**
Damp grassland provides shelter and nest material.

? DID YOU KNOW?

● Crane designs feature in Japanese marriage ceremonies to symbolize happiness, long life and fidelity.

● In feudal Japan, Shogun leaders let only one crane be killed a year, as a gift for the Emperor's feast.

BREEDING

▲ **TENDER TWO-STEP**
Cranes form and strengthen lifelong pair bond with dances.

Cranes pair for life, reinforcing the bond with dancing. Flocks disperse early in the breeding season and pairs take up their nesting territories, which may be several miles apart.

Nest-building is a joint effort. The male brings plant stems to his mate, who builds a mound 12" high and 5' across. Nests are usually sited in, or at the edge of, shallow water.

Both sexes incubate and rear the chicks, which leave the nest in three days of hatching. The adults are diligent parents, bringing tiny morsels, such as insects, to their chicks. They supply food for several months, until the chicks are capable of feeding themselves.

▼ **SOFT START**
Tawny down is replaced by feathers after eight weeks.

CONSERVATION

The Japanese crane's plight began in 1868, when westernization followed the ousting of the Shoguns (feudal governors who had long protected the tancho). Hunting controls disappeared; by the early 1900s European hunters shot the crane into presumed extinction. When 20 survivors were discovered in 1924, rescue efforts began; the crane's population now stands at about 1,800 birds. It's still listed as vulnerable and, although cherished by the Japanese nation, remains threatened by habitat loss, pesticide use and fires.

BEHAVIOR

The crane dances almost year-round and at any age; its antics play a key role in almost every aspect of its life. Its routines are used to attract mates, stake out nesting territories, greet other members of the flock and warn of possible danger. Frequent and spectacular dances are performed on snow-covered fields in late winter, when a solo display may inspire a flock into a frenzy.

During their autumn migration, the Asian mainland flocks soar high or fly in V-formation, with a characteristic wing action in which the powerful downbeat is followed by a quick upward flick. The Japanese crane becomes flightless during the summer molt of its wing feathers; during this time it hides in thick cover while the new quills grow and spread into feathers.

TAKE YOUR PARTNER!

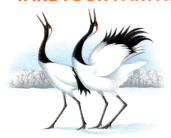   

❶ Duetting…
Graceful and lively, the dance follows no set routine, but the birds usually start by lifting their heads and bugling loudly.

❷ Bowing…
One crane curtsies, while its mate accepts the compliment. The bowing bird may hold a stick or reed, which it tosses into the air as it rises again.

❸ Craning…
Some postures reveal why "craning" describes awkward movements of the human neck. Dancing birds also make much play with their wings.

❹ Leaping
Joyous jumps are performed frequently during the display. When dance fever spreads, an entire flock may hop and skip.

FOOD & FEEDING

The crane's powerful bill and long reach of its flexible neck let it exploit a range of food on land and in water. It obtains much of its food by digging in mud, probing deep for grubs, worms and other invertebrates. It also forages for vegetation, roots, seeds and buds.

The crane captures fish, frogs, snakes and flying insects with a rapid jab of its bill. It also snaps up small mammals and ducklings or other young birds in a similar manner. When faced with prey items too large to swallow whole, the crane shakes them vigorously in its bill to break them into more manageable pieces.

◀ **SHARP SPEAR**
The crane uses its long bill to stab at vegetation and small prey in shallow marshes and rivers.

For the cranes resident on the island of Hokkaido, the staple diet switches in winter to corn, provided at special feeding stations by farmers and conservationists.

PROFILE JAPANESE CRANE

The crane's long legs and neck aren't only useful for feeding; they also enhance the statuesque elegance of its dance rituals.

CROWN
The naked patch of red skin on the crown becomes enlarged and more intensely colored in threat or courtship display.

BILL
Long, strong and pointed, the bill is both a dagger and a digger, used for stabbing at fish and animal prey, as well as rooting out food from the soil.

JUVENILE
The immature crane differs from the adult bird in having a brown neck and black tips to the primary (wing-tip) feathers.

WINGS & TAIL
When the wings are folded, the short, white tail is hidden beneath a black bustle, formed by elongated and pointed inner wing feathers, known as *tertials*.

LEGS AND FEET
Long legs are typical of large birds that wade in water. Adult cranes do not swim, but chicks can swim remarkably well.

CREATURE COMPARISONS

All tall and elegant, cranes vary greatly in size. Smallest is the demoiselle crane of Asia and Africa, which stands half the height of the Japanese crane. With its golden topknot, the 3'-tall crowned crane is a striking inhabitant of the African savannah. Up to 4' tall, the sandhill crane lives in Siberia and North America. Like the Japanese crane, it has a red crown.

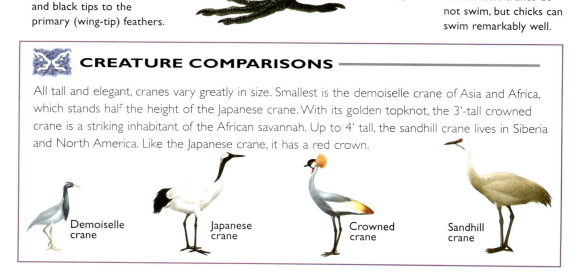

Demoiselle crane • Japanese crane • Crowned crane • Sandhill crane

VITAL STATISTICS

WEIGHT	13–20 lbs.
HEIGHT	4'
WINGSPAN	7'
SEXUAL MATURITY	3–4 years
MATING SEASON	March–May
NUMBER OF EGGS	2
INCUBATION PERIOD	4–5 weeks
FLEDGING PERIOD	11–13 weeks
BREEDING INTERVAL	1 year
TYPICAL DIET	Plants, grain, roots, insects, reptiles and small mammals
LIFESPAN	Up to 25 years

RELATED SPECIES

● The Japanese, or Manchurian, crane is one of 14 crane species; seven are threatened. The rarest (and most closely related to the Japanese) is North America's whooping crane (below), with a population of about 300.

Kakapo

• ORDER •
Psittaciformes

• FAMILY •
Psittacidae

• GENUS & SPECIES •
Strigops habroptilus

KEY FEATURES

- The world's only flightless parrot and, at about 100 times the weight of a budgerigar, the heaviest
- Almost wiped out in its natural habitat by predators introduced by humans, it now survives on a few small islands under strict protection
- Unusual for a parrot, it's almost always solitary

WHERE IN THE WORLD?

Native to New Zealand, but extinct across its former natural range on the mainland; introduced to three nearby islands: Little Barrier Island, Codfish Island and Maud Island

Kakapo 545

LIFECYCLE

Unable to fly, the kakapo relies on camouflage as its defense. Very vulnerable to predators introduced by humans, it's now among the world's rarest birds.

HABITAT

The kakapo was widespread in beech and podocarp (a dwarf conifer) forests of southern New Zealand, but Maoris and, later, European settlers, cleared 75% of this native vegetation. Settlers also introduced "alien" predators, including cats, stoats, weasels and dogs. Grazing animals, such as sheep, also drastically altered its habitat.

In the face of these threats, the kakapo declined until only a handful of birds were left. To ensure their future, these birds were captured and released into safe sites. Remaining birds now live only on a few small islands off the coast of New Zealand's mainland islands.

▲ **NEW ISLAND HOME**
Kakapo habitat on Little Barrier Island, New Zealand.

▼ **GREEN HABITS**
The kakapo is colored to blend in with its habitat.

CONSERVATION

The species was abundant in the south and west of New Zealand's South Island until 1900; fewer than 60 kakapos now survive. In the 1970s and 1980s, the birds remaining on Stewart Island (a stronghold off the south coast of South Island) were moved to remote islands where all introduced predators were exterminated. Relocated populations require supplementary feeding; these islands support few trees that produce the high-protein fruits needed for successful breeding.

DID YOU KNOW?

● An old English name for the kakapo is "owl parrot," a reference to its owl-like face and nocturnal habits.

● Unlike many species of parrot, it has so far proved impossible to create the right conditions to keep and breed kakapos in captivity. The species can only survive in the wild.

● Unusual for a bird, the kakapo has a strong and distinctive smell and may scent-mark its surroundings like a mammal.

BEHAVIOR

Unlike other parrots, the kakapo is solitary — except in the breeding season. Each bird occupies a territory of about 0.2 sq. mile, spending the day under shrubs, in burrows under tree-roots or in rock crevices. At dusk, the almost entirely nocturnal kakapo feeds.

The kakapo can't fly, but is an agile climber, using its feet and bill while balancing with outspread wings. It moves with surprising speed downhill by running with a clumsy gait and gliding for short distances. When kakapos move to higher ground to feed or breed, they tend to follow the same routes; over the years, their movements wear away tracks through the undergrowth.

FOOD & FEEDING

Strictly vegetarian, the kakapo feeds on a variety of fruits, seeds, shoots, leaves, roots, moss, fungi and tubers. It searches for food on the ground, climbs into the lower branches of shrubs and trees or digs up roots and tubers with its sturdy bill. With an action similar to that of a pair of gardening shears, the bill can slice through tough stems.

Kakapos chew food with their bill, unlike the great majority of birds, which use a muscular part of the stomach, called the *gizzard*, to grind their food. The kakapo's bill, though, is equipped with a series of ridges on the inside of the upper mandible. These work against the tongue and lower mandible to shred tough, fibrous food.

◀ **LONER**
Most parrots are sociable, but not the kakapo.

▼ **UNDERCOVER**
During the day, kakapos rest under rocks, roots or shrubs.

▲ **FEET FIRST**
The kakapo often grasps large or tough food items with one of its feet.

BREEDING

The kakapo doesn't breed every year; its normal diet is too poor in nutrients. High-protein seeds and fruit are needed before males can perform energetic courtship rituals and females come into breeding condition. Good crops are produced only every few years, and the kakapo is restricted to breeding at these times.

▲ **SOUND SYSTEM**
A male's shallow "bowl" amplifies his loud calls in the breeding season.

When breeding, males share a network of tracks leading to shallow "bowls" in the ground, set on prominent ridges, forming a group courtship arena, or *lek*. Each male occupies his own bowl and "booms" continuously for 6–8 hours every night from December to March. Females travel several miles to visit the lek, choose a male and mate.

A female builds a nest, incubates the eggs and raises the offspring by herself. They fledge at around 10–12 weeks.

BIG BOOMER

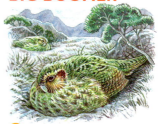

① **Boom in the gloom...**
Males gather at a traditional courtship arena at night. Sitting in their bowls, they boom loudly, as if blowing over the top of bottles.

② **Dancing partner...**
Rival males try to get the females' attention by performing a dancing display. Most females mate with dominant males.

③ **Sitting it out...**
The female builds a simple nest and incubates her 2–3 white eggs. She receives no help from the male in these tasks...

④ **Single parent**
...and for the first 4 weeks, broods the chicks all day, feeding them by night. By 8 weeks, she visits only once or twice a night.

Profile Kakapo

As large as a domestic cat and flightless, the kakapo is an owl-like parrot that is adapted to earn a living on the ground, under cover of the night.

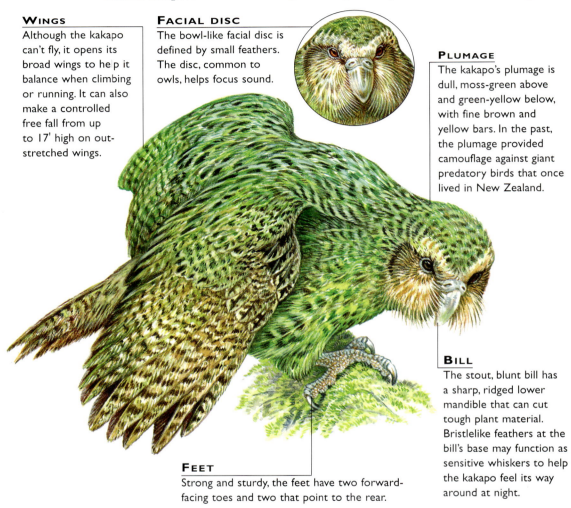

WINGS
Although the kakapo can't fly, it opens its broad wings to help it balance when climbing or running. It can also make a controlled free fall from up to 17' high on outstretched wings.

FACIAL DISC
The bowl-like facial disc is defined by small feathers. The disc, common to owls, helps focus sound.

PLUMAGE
The kakapo's plumage is dull, moss-green above and green-yellow below, with fine brown and yellow bars. In the past, the plumage provided camouflage against giant predatory birds that once lived in New Zealand.

BILL
The stout, blunt bill has a sharp, ridged lower mandible that can cut tough plant material. Bristlelike feathers at the bill's base may function as sensitive whiskers to help the kakapo feel its way around at night.

FEET
Strong and sturdy, the feet have two forward-facing toes and two that point to the rear.

VITAL STATISTICS

WEIGHT	Male 5.5 lbs.; female 4 lbs.
LENGTH	2'
SEXUAL MATURITY	6–8 years
BREEDING SEASON	December –May
NUMBER OF EGGS	2–4
INCUBATION PERIOD	30 days
FLEDGING PERIOD	10–12 weeks
BREEDING INTERVAL	3–5 years; "relocated" birds breed more frequently
TYPICAL DIET	Fruits, seeds, leaves, stems and roots
LIFESPAN	At least 20 years; some reach 30–40

CREATURE COMPARISONS

The kakapo resembles another large parrot native to New Zealand: the kea (*Nestor notabilis*). But the kea is a powerful and agile flier; its bill is much larger than the kakapo's and is strongly hooked with a long, sharply pointed upper mandible. The kea isn't exclusively vegetarian; it uses its bill to dig up burrowing invertebrates and as a meat hook to tear into carrion — especially sheep carcasses.

Kakapo   Kea

RELATED SPECIES

● The kakapo's closest relatives are probably the ground parrot and the little-known night parrot of Australia. There are six parrot species native to New Zealand. The red-crowned parakeet (below) is the only one to occur elsewhere: on Norfolk Island and New Caledonia.

Kea

• ORDER •	• FAMILY •	• GENUS & SPECIES •
Psittaciformes	*Psittacidae*	*Nestor notabilis*

KEY FEATURES

● Labeled a sheep killer, it occasionally feeds on meat but is mainly a vegetarian

● Its far-reaching call echoes from the swirling mists of the mountain tops high in the New Zealand Alps

● It's the only parrot in the world that can survive in snow; frequently remains in high altitudes throughout the winter

WHERE IN THE WORLD?

Limited to the southern portion of the island of New Zealand; found only on South Island, from northwestern city of Nelson through the regions of Marlborough and Fiordland

LIFECYCLE

The multitalented kea shows both brains and brawn in its daily activities, especially in its exploitation of a wide range of available food sources.

HABITAT

▲ **PERFECT PANORAMA**
From the snow-capped mountains to the shining sea, the kea's New Zealand home offers an unequaled island view.

Today, the kea is limited to South Island, the largest of New Zealand's islands. The sturdy kea is at home in the trees or on the ground and is found from just above sea level up to 7,000' in the New Zealand Alps. It prefers higher altitudes, living mainly in the high alpine basins and steep valleys. It is the only parrot in the world that can survive in the snow, frequently remaining in the high altitudes throughout the winter. But it often descends to the lowland river flats, especially on the western side of the southern Alps. The mountain flax plant is common throughout the kea's range, especially on exposed, windswept hillsides. Its flowers provide a sweet treat throughout the summer. During tourist season, the kea frequents camp grounds and ski resorts, looking for food.

CONSERVATION

Since kea scavenge on sheep, which are of economic importance in New Zealand, the birds have unfairly earned a bad reputation. Thousands of kea were killed prior to 1970, but in 1986 the bird, with its near threatened status, received full governmental protection. There are currently about 5,000 kea.

BREEDING

The kea lives in loose flocks, sometimes numbering up to 50 birds, but only a few males and females breed. A small percentage of males, occasionally only 10%, mate in any one season; however, a single male kea may occasionally have several female breeding partners. The pair strengthens their bond by mutual preening. Their nest is made of twigs, grasses, moss and lichens, and sits on the ground — under a log, in a cavity or at the base of a rock outcrop within the forest. Only the female incubates the 2–4 white eggs. The male continues to feed the female after the white down-covered chicks hatch, and brings food for the chicks. After a few weeks, the female finally leaves the nest, and forages for food for the developing nestlings. The male remains with the chicks while the female is gone. Even after the young fledge at 13–14 weeks of age, the parents remain close, accompanying the young for another month to six weeks.

▶ **GROWING UP**
With each molt, the juvenile looks more and more like its parents; it reaches sexual maturity at about 3 years of age.

BEHAVIOR

In flight, the kea announces its arrival with a loud call, "keee-aa." Nonterritorial, the kea moves from place to place, searching for food from snowy ski resorts to flower-filled valleys. Instead of wasting energy by flying, the kea climbs like a monkey to reach berries and buds. The parrot proficiently anchors its bill, then quickly maneuvers its feet.

DID YOU KNOW?

● The playful kea has been observed "sledding" down the A-frame roof of an alpine resort on its back.

● The kea is a member of the *Psittacidae* family, which has the largest number of threatened species of any bird family; about 90 species are at risk.

550 Kea

FOOD & FEEDING

The kea's menu varies from day to day and season to season — it may include anything from sweet fruits to carrion. The kea's selected staples include berries, buds, shoots and roots. It spends much of its time foraging on the ground, digging for roots and underground insects with its trowel-like beak. The bird also delicately removes nectar from flowers in the spring and summer with its comblike tongue. In addition, the kea eats berries and fruits and maneuvers food from bill to foot and back to the bill again with finesse.

Like all parrots, the kea is a very opportunistic bird, and does not pass up the chance to exploit New Zealand's approximately 50 million sheep. It mainly scavenges carrion, but occasionally kills sick or elderly sheep if food is scarce in the winter. The powerful bill, strong enough to tear through sheet metal, is used to rip flesh, much like an eagle's beak.

ANNUAL FARE

❶ November…
During the summer, the kea laps up nectar from the common mountain flax plant.

❷ January…
Late in the season, the kea dexterously feeds on snow totara berries, expertly coordinating its feet, bill and tongue.

❸ March…
As winter approaches and temperatures fall, sick and elderly sheep die, and the opportunistic kea feeds on the carrion.

❹ May
Young kea frequent ski resorts in the cold, snowy winter and scavenge for feasts of food scraps in open refuse bins.

▼ STRONG BOND
The pair is the basic social unit.

Profile   KEA

The green kea cleverly manipulates food with its beak and feet, climbs like a monkey and shows off its brilliant colors.

IN FLIGHT
The kea's olive-green color subtly blends with the foliage, but parrots in transit are unmistakable. The reddish-orange underwings and rump flash in the sunlight, making it easy for fellow flock members to follow in flight.

BILL
Measuring up to 2" from base to tip, the upper mandible functions as a trowel and is slightly shorter in females. The bill manipulates and prepares food, but can also help pull the kea up when climbing.

FEET
The toes form an X, with two toes forward and two toes projecting behind. This allows the kea to grasp branches tightly and also carefully manipulate food to the bill.

VITAL STATISTICS

Weight	Male up to 2.2 lbs.; female up to 1.8 lbs.
Length	Up to 18"
Wingspan	Unknown
Sexual Maturity	Over 3 years
Breeding Season	July–March
Number of Eggs	2–4
Incubation Period	About 24 days
Fledging Period	90–100 days
Breeding Interval	1 year
Typical Diet	Berries, shoots, roots, seeds, nectar, insects, larvae, carrion
Lifespan	About 20 years

RELATED SPECIES

● Along with the kaka, *Nestor meridionalis*, the kea is one of two species in the genus *Nestor*. They join over 300 species in the parrot family, *Psittacidae*. More adept at climbing than most birds, the family members include amazons, cockatiels, lorakeets and rosellas. Also found on New Zealand is the kakapo, *Strigops habroptilus*, the world's only flightless parrot.

CREATURE COMPARISONS

The 18" long olive-brown kaka *(Nestor meridionalis)* is the same length as the kea but, at about 1.3 lbs., weighs one-third less. The kaka is more widespread on New Zealand than the kea and inhabits lower-altitude forests on both North and South Islands. Like the kea, the kaka includes berries, nectar and insects in its diet. But the kaka also chisels away bark to reach the larvae of Kanuka longhorn beetles and licks the honeydew secretions of scale insects.

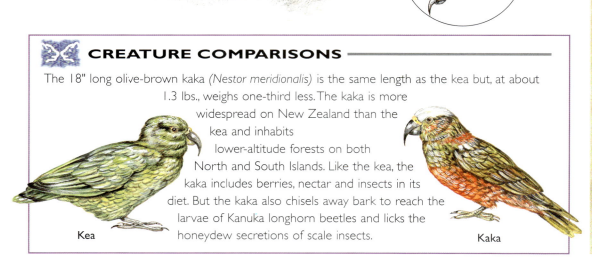

Kea   Kaka

KING PENGUIN

• ORDER •
Sphenisciformes

• FAMILY •
Spheniscidae

• GENUS & SPECIES •
Aptenodytes patagonica

KEY FEATURES

● Holds record among all birds for longest fledging period (up to 13 months)

● Can distinguish its chick's call out of thousands when returning from the sea with food

● Survives the intense cold with the help of increased metabolism, a thick layer of blubber beneath the skin and specialized feathers

WHERE IN THE WORLD?

Found on at least eight sub-Antarctic islands, including Marion and Macquarie Islands, in the southern Atlantic and Pacific Oceans between 46-55° S

LIFECYCLE

A black-and-white blanket forms over sub-Antarctic islands as king penguins cluster during incubation. The mass of huddled juveniles resembles a giant fuzzy carpet.

HABITAT

The king penguin spends most of its time in the icy waters of the Antarctic; penguins gather on land mainly to reproduce and molt on sub-Antarctic islands such as Crozet and Macquarie. Penguin colonies of hundreds of thousands huddle together to combat the intense cold. Natural enemies in these frigid temperatures include leopard seals and killer whales. Penguins avoid these predators with countershading: from above they blend with the jet-black depths, from below with the white ice.

▲ A CHILL IN THE AIR
Warmer than the Antarctic, sub-Antarctic islands are still intensely cold.

BREEDING

King penguins remain in colonies throughout the year, but always return to their ancestral nesting sites to lay their eggs and rear their young. The penguin builds no nest. Instead, it balances a single egg on top of its feet, bending over so that a skin fold covers and warms the egg. Parents take turns incubating the egg. Upon hatching, the soft, downy chick remains on its

parent's feet for about 35–40 days until it grows too large to be warmed. When both parents go off to sea in order to feed their chick's growing appetite, chicks gather together in large creches that help keep them warm and protected from the skua, a predatory bird. Returning parents find their chick by listening for its call, which is literally unique among thousands. It enables parents to identify and feed their own chick. The egg-laying and chick-rearing process takes about 15 months, which is why king penguins breed only twice in three years.

CONSERVATION

The king penguin, once slaughtered for its blubber and skin, was eliminated from the Falkland Islands in the 1870s. It took about 50 years for penguins from major colonies elsewhere to find their way to South Georgia in the Falklands and occupy rookeries there. King penguin populations are currently stable.

FROM EGG TO ADOLESCENT

❶ Getting ready...
"Wall-to-wall" on muddy flats while incubating their egg, king penguins keep their neighbors in place by the jab of a beak.

❷ New arrival...
The female has returned to feed some regurgitated fish to its 2-week-old. The male now goes out to sea for a meal.

❸ Baby food...
Groups of chicks, called creches, provide warmth and protection when both parents go to search for food.

❹ Growing up
A 10-month-old chases its mother, begging for more food. It will join the creche until both parents return with more food.

FOOD & FEEDING

As Antarctic ice breaks up during the longer summer days, tiny floating plants, known as phytoplankton, grow rapidly. These plants supply nutrients for zooplankton and the shrimplike krill, a staple of the king penguin's diet. Krill swim in schools in the top 150' of the sea, a depth easily reached by the penguin. Diving an average of 160' with a maximum dive of 800', the penguin seems to "fly" through the water in pursuit of prey, which includes lanternfish and squid. The king penguin can ingest larger food than other penguins since the span between the tips of its opened bill is greater.

◀ **A RAVENOUS CHICK**
The chick's blubber serves as an energy source to cope with the irregular feedings from its parents.

BEHAVIOR

▲ **MY CHILD'S CALL**
Parents can distinguish their own chick's call among thousands when returning from feeding trips at sea.

Gregarious king penguins flock together on land and in the sea. They waddle comically, tilting their heads to and fro, focusing with one eye then the other. Since they glide faster than they walk, penguins often toboggan toward their nesting sites, following a voluntary leader.

At sea, the paddlelike wings propel the penguin, while the legs, feet and tail provide steerage. The penguin leaps through the air every few yards, clearing the surface mainly to breathe. Leaping also keeps it from becoming easy prey. Insulation alone is insufficient to maintain a safe body temperature in the water, so the penguin's metabolism increases during its dives. The penguins return to land to molt. During molting, islands are covered in feathers, as the penguin loses them in large patches while new ones appear. The new feathers will not become waterproof for several weeks. As a result, the penguin is unable to dive in the water in search of food and often loses 30% or more of its bodyweight.

 DID YOU KNOW?

- Early Antarctic explorers thought the brown, down-covered king penguin chicks were a totally separate species and called them the woolly penguins.

- The name "penguin" was originally given to the now extinct great auk, a large, flightless, black-and-white bird with an extremely upright stance.

- The egg has a chalky-white surface that can easily be removed. Underneath is a harder layer with a pale-green color that intensifies during incubation.

PROFILE KING PENGUIN

A flightless bird, the king penguin is adapted for aquatic existence and is able to "fly" underwater with great speed and agility.

FEATHERS
Small muscles attached to each feather allow the angle between the feather and the king penguin's body to change. On land, feathers are held erect, trapping an insulating layer of air next to the skin. In water, feathers are flattened, forming a watertight barrier for the skin and downy underfeathers.

BILL

The bill, with its bright yellow-orange lower beak, has a razor-sharp cutting edge. The tongue is covered with spiky, inward-pointing spines that grip and prevent the escape of captured fish.

FEET

Strong webbed feet, tucked behind the body when swimming, add to the streamlined shape of the penguin in the water. The three hooked toes grip slippery rocks and ice when the penguin waddles on land.

WING
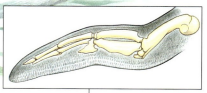
Powerful muscles attach the heavy-boned, paddlelike wings to the sternum.

VITAL STATISTICS

WEIGHT	30–40 lbs.
LENGTH	36–38"
SEXUAL MATURITY	3–8 years, usually 5–6
BREEDING SEASON	November–April; lays egg early one season, late the following season
NUMBER OF EGGS	1
INCUBATION PERIOD	54 days
FLEDGING PERIOD	10–13 months
BREEDING INTERVAL	Breeds twice in 3 years
TYPICAL DIET	Fish, squid and krill
LIFESPAN	17–20 years

RELATED SPECIES

● Penguins are the largest group of flightless birds, with 17 species in six genera. While most species have black backs, the king, emperor and little blue penguin have blue-gray backs. The low mortality rate in the king penguin's genus, *Aptenodytes*, seems to be linked to the laying of a single egg. King and emperor penguins each lay one egg.

CREATURE COMPARISONS

The gentoo penguin (*Pygoscelis papua*) is a close relative of the king penguin but is considerably smaller at 30". The gentoo's black head and irregular white patch above the eye contrast with the large orange patches on the sides of the king penguin's head and neck. The gentoo's breeding grounds range from the Falklands to the South Georgia and Kerguelen Islands, much like the king penguin. Unlike the king penguin, which makes no nest, the gentoo penguin builds a nest out of bones, feathers, grass and stones. It lays two eggs, as opposed to the king penguin's one, but like the king, both male and female share incubation.

Gentoo penguin   King penguin

King Vulture

• **ORDER** •
Falconiformes

• **FAMILY** •
Cathartidae

• **GENUS & SPECIES** •
Sarcoramphus papa

KEY FEATURES

- A powerful scavenger that soars over the forests of Central and South America
- One of the most strikingly colored and odd-looking of all birds of prey
- Often the first vulture to feed at a large carcass; its strong bill can rip through the toughest hide

WHERE IN THE WORLD?

Ranges from central Mexico through Central America into South America, as far south as northern Argentina; also found on the island of Trinidad in the Caribbean

LIFECYCLE

The king vulture often relies on other vulture species to find a meal. But at a carcass, it lives up to its name when others stand back, giving way to the power of its mighty bill.

HABITAT

The king vulture is a bird of the lowland tropics, but occasionally is found up to 8,250' on the slopes of the Andes mountains. It prefers to fly over large, undisturbed areas of forest, but may hunt on open savannah and grassland, although usually only where there is woodland close by. It's at home in the unbroken rainforests of the Amazon as well as in the dry, deciduous forests of parts of Central America, Venezuela, eastern Brazil, Paraguay, Bolivia and northern Argentina.

▲ **LONER**
The king vulture is at home in the deep gloom of dense rainforest.

FOOD & FEEDING

The king vulture is a scavenger, feeding on carcasses of a wide range of wild and domestic animals. Soaring high, it uses its superb eyesight to spot carrion far below in grassland or forest clearings. But above dense forest, it relies on other vultures to guide it to food. Unlike the king vulture — and most other birds — the turkey vulture and two species of yellow-headed vulture, which share the king vulture's range, have a keen sense of smell and use it while soaring at low level to trace carcasses other species can't see. The king vulture takes advantage of this, often shadowing the smaller vultures. When they drop to feed, it follows close behind. It dominates the others at a carcass, but may tolerate their presence while it feeds.

The king vulture has a stronger bill than its relatives, and can even rip into the "armor" of an armadillo (above right). This sometimes benefits the smaller vultures as they can't break into such carcasses themselves.

FIT FOR A KING

❶ Search...
Soaring high, a king vulture spots turkey vultures below as they fly in circles — a sign that they've scented food hidden in the forest.

BREEDING

Courtship begins in March with a pair of vultures opening and closing their wings and lowering their heads to show off their crowns.

Like other New World vultures, king vultures don't build a nest. The female lays her one white egg in a hollow tree stump or on the ground, occasionally on cliffs, and sometimes in the old nests of other birds. The egg is incubated for 8 weeks, in some pairs by both sexes; in others, only by the female. The chick (covered in white down) and the female are fed on regurgitated carrion by the male. Later, the female helps feed her offspring.

▲ **PRINCE**
This youngster has yet to grow full adult plumage.

The chick regularly wanders from the nest as it develops and fledges in three months. The parents keep a close watch on their young for possibly as long as two years, before it leaves and becomes independent.

DID YOU KNOW?

● The king vulture's species name, *papa*, is the Latin word for bishop and relates to the bird's similar appearance to the finery worn by a bishop.

● In Brazil, the turkey vulture — the frequent feeding partner of the king vulture — is called the minister vulture, reflecting its sometimes privileged, but subordinate, position.

● Like its relatives the storks, the king vulture shoots its white, liquid droppings onto its legs to cool itself in hot weather.

② Follow...
The king vulture follows the turkey vultures to the carcass of a river dolphin that washed up on the bank and drives them away.

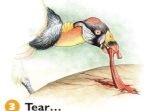

③ Tear...
The king vulture soon rips the dolphin open and begins feeding on the skin and tougher parts while the other vultures look on.

④ Full
When the king vulture has eaten its fill, the turkey vultures draw close, waiting to feed on the softer parts left behind.

CONSERVATION

It's difficult to estimate the king vulture's population, as the bird is rarely seen, spending much of its time in the forest canopy or soaring high in the sky. The species doesn't appear to be threatened and may even benefit from the growing number of cattle ranches, provided it has access to patches of undisturbed forest for breeding. But it suffers when forests are cleared or where populations of large mammals are reduced.

BEHAVIOR

The king vulture isn't very social and is usually found singly, in pairs or in family groups of two parents and a youngster. It spends its time soaring or perching on treetops or resting on river sandbanks. It's a silent bird; along with its relatives, it lacks the *syrinx* (voicebox) of most birds and can make only weak hissing sounds.

The unusual pattern of brightly colored bare skin on the king vulture's head probably helps it recognize members of its own species as well as advertise an individual's status.

▼ **MORNING SPRUCE-UP**
The sun's heat straightens feathers bent by hours spent soaring.

▲ **CLEAN-SHAVEN**
A featherless head is easy to clean after feeding.

PROFILE: KING VULTURE

A powerful build and intimidating bill enable the colorful king vulture to dominate other South American vultures and rip into tough carcasses.

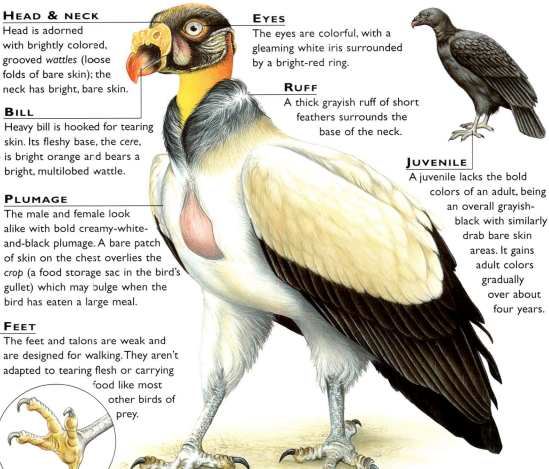

HEAD & NECK
Head is adorned with brightly colored, grooved *wattles* (loose folds of bare skin); the neck has bright, bare skin.

BILL
Heavy bill is hooked for tearing skin. Its fleshy base, the *cere*, is bright orange and bears a bright, multilobed wattle.

PLUMAGE
The male and female look alike with bold creamy-white-and-black plumage. A bare patch of skin on the chest overlies the *crop* (a food storage sac in the bird's gullet) which may bulge when the bird has eaten a large meal.

FEET
The feet and talons are weak and are designed for walking. They aren't adapted to tearing flesh or carrying food like most other birds of prey.

EYES
The eyes are colorful, with a gleaming white iris surrounded by a bright-red ring.

RUFF
A thick grayish ruff of short feathers surrounds the base of the neck.

JUVENILE
A juvenile lacks the bold colors of an adult, being an overall grayish-black with similarly drab bare skin areas. It gains adult colors gradually over about four years.

VITAL STATISTICS

WEIGHT	6.5–8 lbs.
LENGTH	2–2.7'
WINGSPAN	6–6.5'
SEXUAL MATURITY	3–4 years
MATING SEASON	March–August
NUMBER OF EGGS	1
INCUBATION PERIOD	53–58 days
FLEDGING PERIOD	About 3 months
BREEDING INTERVAL	1 year
TYPICAL DIET	Carrion of a wide variety of species
LIFESPAN	Unknown

CREATURE COMPARISONS

Two other species of vulture from the New World are even larger than the king vulture: the California condor, *Gymnogyps californianus*, and Andean condor, *Vultur gryphus*. The latter is in the same family as the king vulture and is the world's largest bird of prey. The Andean condor can be four times heavier than the king vulture; large males weigh up to 26 lbs. It has an overall length of up to 4' and a wingspan of up to 10.5'. Like the king vulture, it has a bare neck to prevent soiling when feeding on carrion. The Andean condor has also been known to kill sickly or dying animals.

King Vulture

Andean condor

RELATED SPECIES

● The king vulture is one of 7 species of New World vulture in the *Cathartidae* family, which includes the American black vulture, California and Andean condors, the lesser and greater yellow-headed vultures and the turkey vulture. New World vultures, classified in the bird-of-prey order *Falconiformes,* are now thought to be more closely related to storks, family *Ciconiidae,* in the order *Ciconiiformes.*

Kiwis

• ORDER •
Struthioniformes

• FAMILY •
Apterygidae

• GENUS & SPECIES •
Apteryx sp.

KEY FEATURES

- Nocturnal, flightless birds that spend the day sheltering in dense cover or hollow logs
- Use long, probing bills and a highly developed sense of smell to search out underground prey
- Produce enormous eggs that are incubated for longer than any other species of bird

WHERE IN THE WORLD?

The brown kiwi is found only on North, South and Stewart Islands, New Zealand; the little spotted kiwi is on four offshore islands; the great spotted kiwi is found on South Island

LIFECYCLE

Kiwis are extremely secretive birds, spending their days hidden in dense vegetation and emerging only under cover of darkness to probe for food buried in the ground.

HABITAT

The three species of kiwi are found in a variety of habitats across New Zealand. Preferred natural habitat is among the wet forests of *podocarps* (evergreen coniferous shrubs) and hardwoods, but they're also found in dry, open forests, scrubland and meadows surrounded by woodland. Because of clearance of native forests in New Zealand (which is now controlled), kiwis are also found in nonnative pine forest plantations and even on agricultural land.

Little is known about kiwis' precise habitat requirements. But soil texture appears to be an important factor; it has to be soft enough for kiwis to probe for food, yet firm and supportive enough to allow safe excavation of nest sites. Soil must also be rich to attract worms and small invertebrates on which kiwis feed.

▼ **UNDERCOVER**
Flightless, kiwis need thick vegetation for cover.

DID YOU KNOW?

- Kiwis have poor eyesight; they can see only about 3' in daylight and 7' in the dark.
- While incubating eggs, male kiwis lose as much as a fifth of their bodyweight.

BREEDING

Once a kiwi chooses a breeding partner, the pair stays together for life. After mating, a female lays one or two eggs in vegetation, a hollow log or between tree roots. A second egg may be laid up to 30 days after the first. Eggs are huge, weighing up to a fifth of the female's bodyweight. The male takes over, incubating the eggs for three months (the longest incubation period of any bird). Hatching may take three days; in a week, the chicks are strong enough to leave the nest to search for food.

◀▶ **ON THE NEST**
A newly hatched brown kiwi *(left)* and a week-old great spotted kiwi *(right)*.

KIWI FRUIT

❶ **Preparation...**
A little spotted kiwi uses its strong legs and feet to excavate a nest hole among the tree roots.

❷ **Big egg...**
The female then lays her enormous egg, the largest of any bird in the world relative to her bodyweight.

CONSERVATION

Full protection was given to kiwis in 1921, but by then the birds had suffered from relentless hunting for food and feathers, and by the destruction of their habitat. Deforestation continues, but conservationists are translocating kiwis to new areas before their forests are cut down and holding some birds in captivity to set up breeding programs.

BEHAVIOR

Many of the kiwis' habits are more akin to those of mammals than birds. They live in pairs all year and keep in contact within a home range by using calls. They're also aggressively territorial, and although they chase away intruding kiwis, they also use droppings to mark out their territories as many mammals do.

Kiwis are nocturnal; because they're flightless, the dark offers safety against predators. In fact, kiwis appear to have a strong dislike of daylight and when roosting, bury their heads beneath the feathers of their rudimentary wings.

 ## FOOD & FEEDING

Kiwis rely on their highly developed senses of smell and hearing to find food during their nighttime foraging. These nocturnal birds scratch through leaf litter and soil with their powerful claws for earthworms, millipedes, beetles, insect larvae, crickets and spiders. Kiwis also employ their long, sensitive bills to target prey, peppering the ground with holes up to 0.4" wide and 6" deep. Seeds and berries are also eaten; kiwis pick them off the forest floor with a tweezerlike bill action. As they feed, kiwis make a snuffling sound as they breathe in to pick up the scent of prey and as they breathe out, possibly to clear dirt from the nostrils after probing in the soil.

When food is plentiful, kiwis amass fat (may total one-third their weight), which they draw on during food shortages.

▼ **BILL OF FARE**
Kiwis can smell prey hidden beneath the surface.

③ Big break…
The chick hatches after three months' incubation by the male. It breaks through the shell with its feet.

④ Freedom
Both parents feed the chick. In a week, it leaves the nest and is independent at 14–20 days.

▼ **NIGHT BIRD**
Kiwis leave their shelters only after dusk.

Kiwis 563

PROFILE KIWIS

Oddities of the bird world, the flightless kiwis have no visible wings and plumage that looks more like a coat of long hair than feathers.

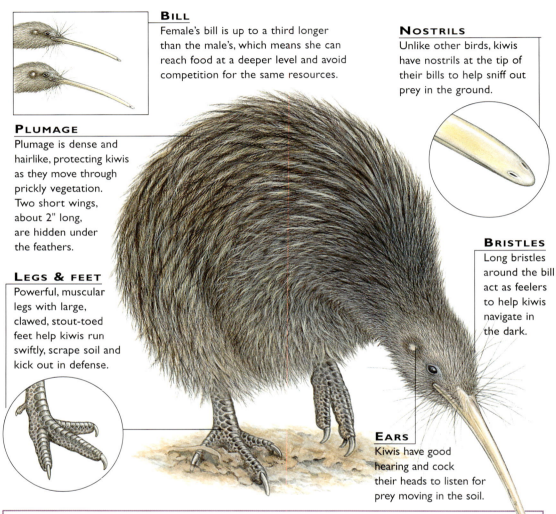

BILL
Female's bill is up to a third longer than the male's, which means she can reach food at a deeper level and avoid competition for the same resources.

NOSTRILS
Unlike other birds, kiwis have nostrils at the tip of their bills to help sniff out prey in the ground.

PLUMAGE
Plumage is dense and hairlike, protecting kiwis as they move through prickly vegetation. Two short wings, about 2" long, are hidden under the feathers.

LEGS & FEET
Powerful, muscular legs with large, clawed, stout-toed feet help kiwis run swiftly, scrape soil and kick out in defense.

BRISTLES
Long bristles around the bill act as feelers to help kiwis navigate in the dark.

EARS
Kiwis have good hearing and cock their heads to listen for prey moving in the soil.

VITAL STATISTICS

WEIGHT	2–8 lbs.; female heavier than male
LENGTH	1–2'
WINGSPAN	1.5–2"
SEXUAL MATURITY	5–6 years
BREEDING SEASON	August–January
NUMBER OF EGGS	1 or 2
INCUBATION PERIOD	71–84 days
FLEDGING PERIOD	14–20 days
BREEDING INTERVAL	1 year
TYPICAL DIET	Insects, worms and berries
LIFESPAN	Unknown in wild; 30 years in captivity

RELATED SPECIES

● There are 3 species of kiwi in the *Apteryx* genus (the great spotted, *A. haasti*, little spotted, *A. owenii*, and brown, *A. australis*). All are in the *Struthioniformes* order (flightless birds), which includes the emu, *Dromaius novaehollandiae* (below).

CREATURE COMPARISONS

Size is the principal difference between the three kiwi species. The largest is the brown kiwi; some females weigh 8 lbs. At the other end of the scale is the little spotted kiwi, which weighs only 2 lbs. The great spotted kiwi is slightly larger. The species can also be told apart by their coloring. The brown kiwi is uniform brown, the little spotted kiwi is brown with lighter barring and the great spotted kiwi has an overall brown-streaked appearance.

Brown kiwi Great spotted kiwi Little spotted kiwi

Kori Bustard

• ORDER •	• FAMILY •	• GENUS & SPECIES •
Gruiformes	Otididae	Ardeotis kori

KEY FEATURES

- At nearly 40 lbs. and with a wingspan of nearly 8', the kori bustard is one of the world's heaviest flying birds
- The bustard will breed in an area just after animal herds have moved through, possibly looking for stirred-up insects to feed its young

WHERE IN THE WORLD?

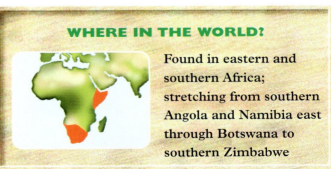

Found in eastern and southern Africa; stretching from southern Angola and Namibia east through Botswana to southern Zimbabwe

LIFECYCLE

Both sexes of the kori are well camouflaged with their cryptic plumage, but the polygamous male invites attention by inflating his neck and ruffling his neck feathers.

HABITAT

The kori bustard prefers flat, arid and mostly open country, generally below 700', with a short herb layer. The bird typically favors locations where the grass is not too long and where stony outcrops are present. It frequents grassland, bushveld, scrubland and savannahs, as well as floodplains, duneland and fossil valleys. The bustard doesn't shy away from man-made habitats, such as wheat fields. During the hot, dry season in Kenya, many birds move into woodland.

▲ **HITCHING A RIDE**
On the plains, a bustard gives a bee eater a ride.

FOOD & FEEDING

Equipped with an all-purpose bill, the kori bustard is an opportunistic feeder. The bustard is well known for taking advantage of swarming locusts and grasshoppers, and after bush fires has been known to feed on the victims, including dead and dying insects as well as vertebrates, mainly small snakes and lizards. Additionally, the bird exploits a wide range of vegetable and animal resources, such as seeds, berries, bulbs, Acacia gum, snails, rodents and small birds. In fact, research has shown that vegetable matter, including seeds, berries and roots, makes up a surprisingly large proportion of the kori bustard's extremely varied diet.

WINE AND DINE

❶ **Join the crowd...**
The kori is not afraid of larger animals, and actually nests in areas where herds have passed through.

❸ **Salad with dinner...**
The kori bustard's sharp bill can also be used to peck at a small shrub, where it pulls off leaves.

? DID YOU KNOW?

● Plant materials consumed by the bustard include the gumlike sap of acacia trees, which may be the basis for the kori bustard's Afrikaans name *Gompou*, or "rubber peacock."

● Prehistoric man seems to have valued the kori bustard; cave drawings and rock engravings feature the bird as game. Currently, the kori bustard features prominently in the dances and songs of Botswana's African bushmen.

BEHAVIOR

The kori bustard male exhibits grand and vibrant displays, most notably its strutting and booming call. During the strutting "balloon display," the male gulps air and inflates his gular pouch, an expanded area of the bird's esophagus, or gullet. The pouch can be inflated to four times its normal size and held in this inflated state for an indefinite period. The bird's "booming" call consists of three pairs of drumlike sounds: *ump-ump, ump-ump, ump-ump*. The kori bustard often associates with herds of large ungulates, such as wildebeest or zebras; the bird feeds on disturbed insects or on insects, such as the dung beetle, that are attracted to the dung piles left behind by these large animals.

▲ **BALLOON DISPLAY**
The male bustard can greatly inflate his neck.

566 Kori Bustard

② Sight and snatch...
Striking out at a dung beetle, the kori bustard uses its bill to snatch the tasty insect.

④ After-dinner drink
A thirsty bustard crouches at a small pool in its grassland home and drinks by filling its bill.

BREEDING

The kori bustard breeds from September to February in South Africa and December through August (depending on the rains) in East Africa. In fact, in East Africa, breeding success is greater when the wet season is longer. Males mate with several females, and while courting, the male walks slowly around the female or stands within 30' of her, bowing with his body tilted forward and neck inflated; the head never reaches below the level of the shoulders. The booming display is performed with a fully inflated neck, the wings drooping, and the tail lowered so as to form a straight line with the wings. The nest is built by the female; it is often a simple scrape on the ground with a thin lining of grass. Nests may be built near rocks or a clump of grass, sometimes in partial shade. The bustard will often breed in an area just after herds have moved through; there, they find fresh insects for their young, which have been stirred up by the mammals. The female usually lays two eggs, but only one in drier years. The incubation period lasts around 25 days, and chicks are capable of flight by 5 weeks of age.

▶ **DAZZLING DANCE**
Bustards perform an elaborate courtship dance.

CONSERVATION

The kori bustard is not globally threatened, but is listed on CITES Appendix II. If undisturbed, the bird can still be common, as it is in parts of Botswana. Around 5,000 individuals inhabit Zimbabwe, but the species is declining there due to habitat destruction, hunting pressure and disturbance. The kori bustard is also threatened in South Africa for the same reasons.

PROFILE: KORI BUSTARD

The kori bustard's large size deters many predators; a walker rather than a runner or flier, it sometimes passes weeks without taking wing at all.

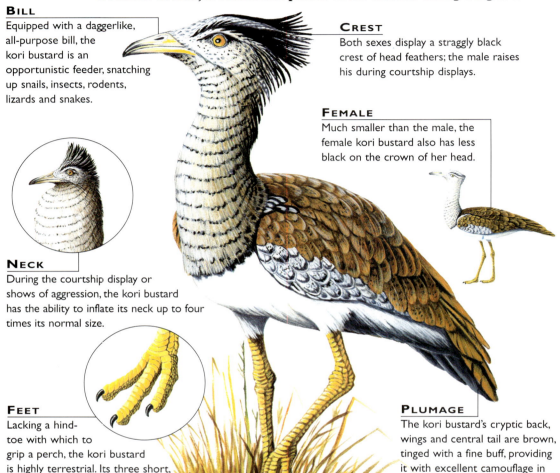

BILL
Equipped with a daggerlike, all-purpose bill, the kori bustard is an opportunistic feeder, snatching up snails, insects, rodents, lizards and snakes.

CREST
Both sexes display a straggly black crest of head feathers; the male raises his during courtship displays.

FEMALE
Much smaller than the male, the female kori bustard also has less black on the crown of her head.

NECK
During the courtship display or shows of aggression, the kori bustard has the ability to inflate its neck up to four times its normal size.

FEET
Lacking a hind-toe with which to grip a perch, the kori bustard is highly terrestrial. Its three short, broad toes are small for the relatively large size of the bird.

PLUMAGE
The kori bustard's cryptic back, wings and central tail are brown, tinged with a fine buff, providing it with excellent camouflage in its habitat; its white wing panels are spotted with black.

VITAL STATISTICS

WEIGHT	About 40 lbs.
LENGTH	4.5'
WINGSPAN	8'
SEXUAL MATURITY	Probably 1 year
BREEDING SEASON	September–February in southern Africa; December through August in East Africa
NUMBER OF EGGS	1–3
INCUBATION PERIOD	25 days
FLEDGING PERIOD	4–5 weeks
BREEDING INTERVAL	1 year
TYPICAL DIET	Seeds, lizards and snails
LIFESPAN	Unknown

CREATURE COMPARISONS

At 47", the Australian bustard (*Ardeotis australia*) is roughly the same size as the kori bustard. Its cryptic plumage is also similar; the back, wings and central tail of the Australian bustard are brown with fine, dark marbling, but its neck spots look more like freckles, in contrast to the kori bustard's bars. Both birds inflate their necks during courtship displays; in addition, the Australian bustard's throat pouch nearly touches the ground in its dramatic "balloon display." The birds' ranges are different: the Australian bustard is found in dense grassland in Australia and New Guinea, while the kori bustard makes its home in the open plains of Africa.

Australian bustard

Kori bustard

RELATED SPECIES

● There are 11 genera and 25 species in the family *Otididae*. The kori bustard is 1 of 4 species in the genus *Ardeotis*, which also contains the Arabian bustard, *A. arabs*, the Great Indian bustard, *A. nigriceps,* and the Australian bustard, *A. australis.* There are a variety of bustard relatives in the order *Gruiformes,* including trumpeters, rails, buttonquails and cranes.

Lammergeier

• ORDER •
Falconiformes

• FAMILY •
Accipitridae

• GENUS & SPECIES •
Gypaetus barbatus

KEY FEATURES

● One of the world's largest birds of prey, this vulture has bold plumage and an impressive "moustache"

● Magnificent in the air, with a striking flight silhouette and the ability to soar effortlessly for hours on end

● Obtains most of its food by scavenging for carrion and breaking bones to expose the nutritious marrow

WHERE IN THE WORLD?

Ranges in central and southwest Asia, from Turkey to China and Mongolia; also found in East and South Africa, the Atlas Mountains of Africa, Corsica, Greece, the Pyrenees and the Arabian Peninsula

LIFECYCLE

The lammergeier is a scarce but legendary hunter and scavenger, renowned not just for its majesty in the air, but also for its spectacular use of gravity to break open bones.

HABITAT

The lammergeier favors wild, rugged mountainscape altitudes of 3,300–9,900', but it's sometimes found on peaks, up to 14,850' that are free of permanent ice. Its common haunts are crags and steep ridges where there are flat, level rocks that can be used as bone-breaking platforms. The lammergeier also occurs in lowland wilderness, including open steppes and tropical plains, exploiting thermals that rise over the latter.

The lammergeier is scarce in Europe, where it is confined to the most remote sites.

▼ **RIDING THE THERMALS**
The lammergeier lives in South Africa's Drakensberg Mountain.

CONSERVATION

The lammergeier is extinct in many areas due to hunting and poisoning. Since 1986 a conservation program has reintroduced 60 birds to the Alps of France, Switzerland, Austria and Italy.

BEHAVIOR

▲ **EYE IN THE SKY**
The lammergeier soars high over its range on thermals and updrafts.

Although a lone hunter, a pair of lammergeirs occupies and carries out most of its feeding in home ranges as large as 120 sq miles. Within this living space, the pair typically uses up to five different roost sites and maintains several alternative eyries, used in rotation from one nesting season to the next.

With apparent ease, a lammergeier may fly over 24 miles in one day while hunting. After a few heavy downstrokes to lift itself from its perch, the bird sets off with its great wings held straight, soaring upward on warm thermals with barely a wingbeat, or gliding parallel to the ground with an occasional flap. At times the lammergeier flies to great heights, but it also spends more time than most vultures in the lower airspace, a few yards from the ground.

To descend, the bird spreads and angles back its wings — rather like a falcon — then glides or swoops, pulling up at the last moment to settle gracefully.

The lammergeier perches with its body held in a more oblique posture than do most vultures, with its head held up and the wingtips free of the tail. Its powerful legs enable it to walk over ground with ease.

▶ **TELL TAIL**
In flight, the lammergeier's tail appears as a distinctive, elongate wedge.

DID YOU KNOW?

● Owing to its prominent beardlike bristles that grow forward along the bill, the lammergeier is sometimes referred to as the "bearded vulture."

● In parts of Asia the lammergeier may visit rural settlements on ritual slaughter days, hoping to snatch scraps from freshly butchered carcasses.

🦴 FOOD & HUNTING

The lammergeier is primarily a scavenger. Despite its size, it usually waits until other raptors, such as vultures, have had their fill before visiting a carcass. It bites clean through small bones, or holds them in its bill and bashes them against rocks.

To extract the marrow from larger bones, this bird repeatedly drops them onto an "ossuary," an outcrop of flat rock. It does the same to crack tortoise shells. Occasionally it attacks sickly goats on rocky slopes, dislodging them by furiously flapping its wings.

▲ **CLEANING UP THE SCRAPS**
The lammergeier eats bones as long as 4" without breaking them. Digestion begins on the bone end in the stomach while the other end is still in the bird's bill.

BONE-BREAKER WITH PATIENCE

❶ A deserted carcass…
The lammergeier returns to a carcass after other scavengers have fed on it; it specializes in hacking into tough skin and bones.

❷ With no meat left…
Grasping a heavy leg bone in its talons, the lammergeier flies to one of its bone-breaking sites. It climbs to a height of 165–264'.

❸ But I'll settle for bones…
As it approaches the spot, the bird dips slightly to increase momentum; it drops the bone, then turns abruptly and follows it down.

❹ If I can break them
Up to 50 drops may be needed before the bone breaks on the rock. Further blows against the hard ground expose the marrow.

BREEDING

Breeding lammergeier pairs stay together for life. Up to three months before breeding, the pair re-establishes their nest site in a niche in a cliff. The birds strengthen their pair-bond with exchanges of food, mutual preening and spectacular aerial dances involving spirals, dives, rolls and twists. In one flight display, the birds chase each other before gripping claws and tumbling to within yards of the ground. The pair refurbishes the nest with wool and dry dung before the female lays her clutch of mottled eggs.

Compared with other large raptors, the male lammergeier plays a major part in caring for the young. Both sexes take turns incubating the clutch and feeding the nestlings; they continue to bring food morsels for a few weeks after the chicks have their flight feathers.

▲ **BLOODLINE**
The adult lammergeier brings food back to the nest for its partner and hungry fledglings.

PROFILE LAMMERGEIER

Massively built, the lammergeier has the long and slender wings of a glider and the strength to lift heavy bones into the sky.

BILL
Long, dark bristles obscure the true size of the lammergeier's hooked bill, which is nearly 3" long from flattened tip to wide base and the same length as the rest of the head.

NECK
Most vultures have a bare head and neck for probing inside carcasses, but the lammergeier has feathers that extend over its throat, forehead and nape.

JUVENILE
The first full plumage of a lammergeier is more subdued in color — gray-brown with dark-brown neck and flight feathers — than that of a mature adult.

FEET
The large feet, strong toes and sharp, curved claws are ideal for lifting and carrying food items. Other vultures can carry only food in the bill.

VITAL STATISTICS

WEIGHT	10–15.5 lbs.
LENGTH	3.5–3.8'
WINGSPAN	8.75–9.3'
SEXUAL MATURITY	5 years
BREEDING SEASON	Varies according to region; January–July in southern Europe
NUMBER OF EGGS	Usually 1 or 2, occasionally 3
INCUBATION PERIOD	55–60 days
FLEDGING PERIOD	100–110 days
BREEDING INTERVAL	1 year
TYPICAL DIET	Hunts small mammals and birds; carrion
LIFESPAN	Unknown

RELATED SPECIES

● Birds of prey form a large order, *Falconiformes*, of 5 families. *Accipitridae*, the biggest family, includes hawks, buzzards, kites, harriers, eagles and Old World vultures. The lammergeier is sole member of its genus, but the family has 13 other vulture species with 8 genera. The 7 species of New World vulture form a separate family, *Cathartidae*, and are not related to their Old World counterparts.

CREATURE COMPARISONS

The bird that most resembles the lammergeier in general shape is the smaller Egyptian vulture (*Neophron percnopterus*), which has a similarly broad range. However, the Egyptian vulture is only two-thirds the size of the lammergeier, with a striking white plumage and a more delicate bill free of bristles. In both species the juvenile has dark plumage, but the lammergeier is larger in size with narrower wing points than the Egyptian vulture. Both are carrion feeders and have been known to use rocks to break open food items. Using its bill, the Egyptian vulture hurls stones against ostrich eggs to get its contents.

Egyptian vulture   Lammergeier

Lappet-faced Vulture

• ORDER •	• FAMILY •	• GENUS & SPECIES •
Falconiformes	*Accipitridae*	*Torgos tracheliotus*

KEY FEATURES

- The lappet-faced vulture is the largest, and among the rarest, of all of the African vultures

- Its lappets — loose folds of skin on its face — look like ears, hence its other common name, African eared vulture

- Has the strongest beak of any vulture, easily tearing into the tough hide of almost any animal

WHERE IN THE WORLD?

In Africa, from the Sahara, east to Ethiopia, south through Kenya, Tanzania and into South Africa, and west to the Namib Desert; also the deserts of Israel and the Arabian Peninsula

LIFECYCLE

Other vultures gather around tough-hided carrion and wait for the lappet-faced vulture, the most powerful of all of the vultures, to arrive and rip the carcass open with its beak.

HABITAT

The lappet-faced vulture prefers the bright sun and warmth of the semi-arid deserts and savannahs of Africa, Israel and the Arabian Peninsula. This massive bird frequents open desert areas with desert scrub; it nests in the shortest trees in the area, commonly thorny acacias. These large, open areas with little cover make it easier for the lappet-faced vulture to spot the dead and dying animals upon which it feeds. Though quite dominant at feeding sites, the lappet-faced vulture prefers to build a solitary nest, away from other breeding pairs.

▼ **WIDE, OPEN SPACES**
The lappet-faced vulture relies on large, open areas for hunting.

 DID YOU KNOW?

- It is thought that some nonindigenous species of plants and trees in the Negav Desert area of Israel were brought there as seeds on the feet of migrating lappet-faced vultures.

- The vulture's bare head is thought to be an adaptation to prevent germs and bacteria, acquired when sticking its head into putrid carcasses, from causing infections. A head covered with feathers is harder to clean thoroughly.

FOOD & FEEDING

The lappet-faced vulture is a voracious eater and is capable of cleaning a carcass down to the bones. When it locates a piece of carrion, it uses its powerful beak to rip holes in the animal's tough hide. The vulture's long neck allows it to probe deep into carrion in search of the large meaty muscles. It then rips off large pieces of meat while holding the carcass with its feet. The largest of all vultures, the lappet-faced is well respected at carrion sites — other animals, including hyenas, are easily driven away when confronted. Due to the scarcity of carrion, especially in the desert areas, the vulture feeds heavily at each sitting, and its crop, an enlargement of the esophagus, can store more than 13 lbs. of food at a time. Once food is digested, the lappet-faced vulture, like all birds of prey, regurgitates pellets of hair and feathers. Although largely a scavenger, it also hunts live game when carrion is scarce. Its prime targets are flamingos, both young and adult, hares, gazelle calves and even locusts and termites.

BEHAVIOR

The lappet-faced vulture is one of the shiest and most solitary of the Old World vultures — except when feeding. Then, the vultures congregate, occasionally gathering in groups of up to 100 birds. Once, 35 lappet-faced vultures were observed surrounding a single dead donkey. The lappet-faced vulture is normally a silent bird but, when gathered around a piece of carrion, it grunts, growls, hisses and yelps. The strongest and most dominant vulture at kill sites, it can easily bully other vultures and even the largest eagle into submission. It usually moves to the outskirts of the feeding group and attacks others by rushing toward them with its head lowered and wings and neck outstretched. The lappet-faced vulture regularly visits water holes, where it washes off its messy face after eating.

▶ **BACK OFF!**
The vulture protects its meal from other animals.

COMMANDER IN CHIEF

1 Needy citizens...
Many vultures are not strong enough to rip through the tough hide of a buffalo, and rely on the lappet-faced vulture to do so.

2 Saving the day...
With the largest and one of the most powerful beaks, the lappet-faced vulture easily tears through the hide of almost any animal.

3 First in line...
The immense lappet-faced vulture commands the scene; it is the first to eat and even takes food from the others.

4 A job well done!
After a meal, the lappet-faced vulture finds a water hole where it can bathe and wash off the mess from its head and neck.

BREEDING

Lappet-faced vultures are solitary nesters and prefer to be far away from other nesting pairs. Intensive nest defense, mate-guarding and courtship-feeding are all part of the mating rituals. Both the male and female lappet-faced vulture work together to build a massive nest of sticks, up to 10' in diameter, usually atop a thorny tree in the open sun. Once the base of the nest is built, they line it with fur from carcasses and grass. During the nest's construction the pair roosts nearby; even when the nest is finished, they will not use the nest until the egg is laid.

▲ **WORKING TOGETHER**
Both members of a breeding pair share in the parental duties.

The female lays one dull-white egg with brown spots and blotches, and the male and female take turns incubating the egg and searching for food. The chick hatches after about 55 days; one parent remains with the chick, while the other scavenges for food. The adults feed the chick regurgitated carrion, including splinters of bone that provide essential calcium. After about 135 days, the young lappet-faced vulture takes its first flight. The adults often return to the same nest for several years.

CONSERVATION

Though not currently on the endangered list, the lappet-faced vulture is declining in southern Africa because of poisoning and shooting, electrocution by high-voltage towers and a shortage of calcium in the diet of chicks. It is listed in Appendix II of CITES, which strictly controls its export out of Africa.

PROFILE: LAPPET-FACED VULTURE

With its strong beak, broad wings and featherless neck and head, the lappet-faced vulture is well suited for finding and feeding on carrion.

IN FLIGHT
Immensely broad wings and widely spaced primary feathers allow the lappet-faced vulture to glide for long periods.

LAPPETS
The lappets are loose folds of skin hanging off the side of the face; featherless, they pick up less of the putrid fluids and flesh that the vulture devours.

EYES
The vulture's eyesight is very keen; it can spot a dead or dying animal from over 1 mile away.

TONGUE & BILL
The rasplike tongue helps grasp and move chunks of meat to the mouth. The large, powerful bill rips holes in carrion.

FEET
The lappet-faced vulture's feet are much weaker than those of other birds of prey and are designed more for running than grasping.

CREATURE COMPARISONS

The white-headed vulture (*Trigonoceps occipitalis*) has a wingspan of 6.5' and length of 2.75', much smaller than those of the lappet-faced vulture. The white-headed is one of the most colorful vultures in Africa; it has a downy, white head, bright orange and blue beak, pink legs and face, and white secondary feathers and tail leading up the middle of the wing. The white-headed vulture is known as "the searcher," since it sets out earlier in the day in search of food than do other vultures. It is often forced to the outskirts of feeding groups when other vulture species arrive and is left to eat the less nutritious scraps.

White-headed vulture

Lappet-faced vulture

VITAL STATISTICS

WEIGHT	12–21 lbs.
LENGTH	3–4'
WINGSPAN	9–9.5'
SEXUAL MATURITY	About 6–9 years
BREEDING SEASON	Varies with location
NUMBER OF EGGS	1–2; usually 1
INCUBATION PERIOD	54–56 days
FLEDGING PERIOD	125–135 days
BREEDING INTERVAL	1 year
TYPICAL DIET	Mostly carrion; also hunts flamingos, hares and insects
LIFESPAN	About 40 years

RELATED SPECIES

● The lappet-faced vulture is 1 of 15 species of Old World vultures in the family *Accipitridae*, found across Africa, Europe and Asia. They are joined by the New World vultures in the family *Cathartidae* in the order *Falconiformes*. But the Old World vultures actually are more closely related to hawks and eagles, also in the family *Accipitridae*, than to the New World vultures.

Laughing Kookaburra

• ORDER •
Coraciiformes

• FAMILY •
Alcedinidae

• GENUS & SPECIES •
Dacelo novaeguinae

KEY FEATURES

● An opportunistic predator of Australia's bush, this crow-sized kingfisher grabs and stabs prey with its bill

● Master reptile killer that thrashes snakes and lizards to death, often after dropping them from high in the air to stun them first

● Extremely noisy, with a vocabulary of loud calls that sound like chuckles and side-splitting laughs

WHERE IN THE WORLD?

Over much of eastern Australia in a broad band (Queensland in the north to Victoria and southwestern South Australia); introduced to a corner of western Australia and Tasmania

LIFECYCLE

A startling array of humanlike "laughs" have contributed to the celebrity of the laughing kookaburra, one of the most conspicuous and charismatic Australian birds.

HABITAT

The kookaburra is found in some of eastern Australia's most arid habitats, especially dry, open eucalyptus forests, but it prefers to be close to a water source. It also occurs in lightly wooded farmland and vast expanses of scrub (the "bush"). It has also adapted well to urban environments and is common in parks and gardens, which it visits to raid bird tables. Living in a sun-drenched climate, the kookaburra needs the shade offered by trees while waiting patiently to swoop on prey.

▼ **HIGH PROFILE**
High perches, especially those close to open spaces, attract kookaburras.

DID YOU KNOW?

● Although not especially skilled at fishing, the laughing kookaburra often snatches fish from shallow ornamental garden ponds.

● The laughing kookaburra is nearly 10 times heavier than the Eurasian kingfisher and 50 times heavier than the African dwarf kingfisher.

FOOD & HUNTING

Although it's a kingfisher, the kookaburra doesn't hunt over water for fish. But its hunting method is similar to other kingfishers'. It mounts an ambush by watching and waiting from a high vantage point and then, when it spots movement on the ground below, dives down to grab prey in its strong bill. Frequently, it launches its attacks from a perch in a tree, although it also makes use of powerlines, especially in built-up areas.

Australia has a large and abundant range of reptiles, especially snakes; these make up a significant part of the kookaburra's diet.

CONSERVATION

The laughing kookaburra is common throughout its range, thriving in towns, cities and natural habitats. Recent estimates based on the density of territories (number of breeding pairs per given area) in a sample of habitats suggest that its population is in excess of 60 million birds. This kookaburra isn't facing any serious threats and, indeed, receives protection under the same Australian law that forbids the trade of all wild animals.

BREEDING

Holes in tree trunks or cavities in dead and rotten wood make ideal kookaburra nest sites. It also takes over vacant nests of tree termites or occupies holes in the walls of buildings. After mating, the female lays two or three white eggs on the bare floor of the nest chamber.

The male and female pair for life and cooperate to rear their brood. They also receive assistance from nonbreeding "helpers" — offspring from a previous year that haven't yet found mates. Helpers may assist for up to four years before attempting to breed themselves, participating in most activities associated with nesting, such as incubating, feeding the young and defending the territory. The young fledge after six weeks, but remain entirely dependent on their parents and helpers for another two months.

A STUNNING PERFORMANCE

1 Swoop to kill…
Scanning the ground from a tree at the edge of a clearing, a kookaburra spots a snake. It silently launches an attack.

2 Snake bite…
Striking quickly, it seizes the snake in its bill and shakes it violently.

3 Backbreaking work…
The kookaburra thrashes it against the hard earth. While the snake is stunned, the kookaburra repeats its shaking, breaking the snake's back.

4 Down in one
Returning to its favorite perch with the dead snake, the kookaburra swallows it whole without fear of injury.

 ## BEHAVIOR

Named after its most famous call, the laughing kookaburra is especially vocal in early morning and after sundown. The "laugh" itself lasts for ten seconds, beginning and ending with a series of deep chuckles. In between is the loud, remarkably humanlike sound.

When calling from a perch, the laughing kookaburra adopts a characteristic posture, with its tail cocked and bill pointing upward.

▼ **HAVING A LAUGH**
The kookaburra makes its range of raucous or throaty calls year-round to advertise territory ownership.

▲ **STICKING TOGETHER**
The kookaburra pairs for life, and both birds share the tasks of maintaining their territory and caring for the eggs and chicks.

Laughing Kookaburra 579

PROFILE   LAUGHING KOOKABURRA

A true giant among kingfishers, the laughing kookaburra's stocky frame and sturdy bill enable it to tackle sizeable, often dangerous prey.

BILL
To cope with a diet of young birds, amphibians and reptiles, its bill is broader and thicker than fish-eating kingfishers'. It's also tipped with a small hook for gripping struggling prey.

BODY
Heavily built, with a huge head relative to its body size, the laughing kookaburra is the largest kingfisher in the world. The female (shown) is larger than the male.

FANNING
When making its laughing call, the kookaburra cocks and fans its tail. This displays the tail's underside, which is brown and white with gray-brown bars.

FOOT
In common with other kingfishers, the kookaburra's feet are relatively small and weak, but, unlike its relatives, the toes aren't fused together.

TAIL
The tail's banding breaks up the bird's outline as it swoops when hunting, helping it take prey by surprise.

CREATURE COMPARISONS

Blue-winged kookaburra — Laughing kookaburra

In parts of eastern Australia, the blue-winged kookaburra (*Dacelo leachii*) occurs alongside the laughing kookaburra. Similar in size, both have a daggerlike bill. The blue-winged kookaburra has paler eyes and a whiter head. Its tail, rump and areas of its wings are blue. Where the ranges of the two overlap, the blue-winged kookaburra is found in damper habitats — in wet forests and tall stands of trees beside watercourses. It has a wider distribution than the laughing kookaburra, across northern Australia and north to southern New Guinea, where it sometimes inhabits coastal mangrove swamps.

VITAL STATISTICS

WEIGHT	11–17 oz.
LENGTH	16–18"
WINGSPAN	20–24"
SEXUAL MATURITY	1 year
BREEDING SEASON	September–December
NUMBER OF EGGS	2 or 3
INCUBATION PERIOD	24–26 days
FLEDGING PERIOD	33–39 days
BREEDING INTERVAL	1 year
TYPICAL DIET	Rodents, frogs, lizards, snakes, insects, earthworms, crayfish, nestling birds
LIFESPAN	6–10 years

RELATED SPECIES

● The kingfisher family *Alcedinidae* belongs to the order *Coraciiformes*, which includes bee-eater, roller and hornbill families. *Alcedinidae* has 87 species of kingfisher; 22 species feed exclusively on fish. Four nonfishing species are in the laughing kookaburra's genus *Dacelo*: large laughing and blue-winged kookaburras and small rufous-bellied and spangled kookaburras.

Lovebirds

• ORDER •	• FAMILY •	• GENUS & SPECIES •
Psittaciformes	Psittacidae	*Agapornis* spp.

KEY FEATURES

- Family of tiny, highly acrobatic parrots

- Each pair spends long periods nestled side by side, nibbling each other's feathers to strengthen their pair bond

- Some species carry nest material buried among their feathers

- Males and females of different species can mate and produce hybrid offspring

WHERE IN THE WORLD?

Found in tropical and subtropical forests in Africa, from Guinea east to Ethiopia, Kenya and Tanzania, and in parts of Mozambique, Zambia, Malawi, Angola and Namibia; one species confined to Madagascar

LIFECYCLE

Lovebirds are named after their "affectionate" habit of billing and preening each other as if in a loving embrace. This behavior reinforces the bonds between the male and female.

HABITAT

▲ **TREETOP JEWELS**
Flock of Fischer's lovebirds in savannah woodland.

Lovebirds are found in tropical and subtropical forests and more open types of woodland. Each species favors slightly different habitats, but more than one species can be found together.

The red-faced lovebird, which has the widest distribution of any lovebird, lives in primary rainforest broken by clearings, secondary forest and wooded plains up to an altitude of 4,950'. Some lovebird species prefer lowland evergreen forests, while the black-winged lovebird inhabits juniper forests in the highlands of Ethiopia.

Even arid or rocky country is a suitable habitat to such species as the peach-faced lovebird, which is found in dry steppes in southwestern Africa; a few others adapt readily to fields. Nyasa and black-cheeked lovebirds are specialized; both are restricted to open stands of the low-growing mopane tree.

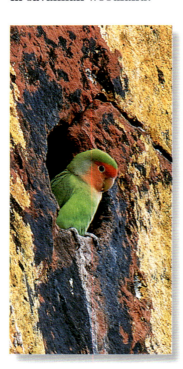

▲ **CLIFF-HANGER**
Peach-faced lovebirds nest in cliffs instead of trees.

 **DID YOU KNOW?**

● Most small birds mate for a few seconds at a time, but lovebirds stay coupled for up to six minutes.

● Up to 25 Nyasa lovebirds may roost together in a single tree hole. They enter one at a time and tail-first.

FOOD & FEEDING

Lovebirds do not travel far, so their home ranges must provide a reliable, year-round supply of food. Most lovebird species live on flowers, seeds, fruit and leaf buds — depending on seasonal availability — which the birds supplement with insects and grubs. Some species, such as the black-collared lovebird, feed mainly on figs in treetops, but others, including the red-faced and gray-headed lovebirds, gather grass seeds at ground level.

▼ **PRETTY IN PINK**
Plants in full flower attract hungry lovebirds.

The lovebirds' superb bill–foot coordination enables them to use its four-clawed feet as a clamp to hold food while the bird breaks up the morsel with its bill. The lovebird may also use a foot to transfer food to its bill. Lovebirds also have a strong tongue to remove seeds from husks.

CONSERVATION

Lovebirds are very popular pets and trapping to supply the cage-bird trade has led to serious declines in the populations of most species. Today, legal protection has helped increase lovebird numbers again, but recovery is a slow process. The black-cheeked lovebird remains endangered and is found only in a 2,400 sq. mile stretch of wooded river valley.

BREEDING

Most lovebirds breed in colonies, but some nest in isolated pairs. Lovebirds make their nests in cliff faces, termite mounds, holes excavated by woodpeckers and nests of swifts, weavers, sparrows and other birds.

The female builds and guards the nest, incubates the eggs for three weeks, then broods the chicks. The male doesn't help her until the chicks are older, when he starts to bring food to them. The young lovebirds are born blind and helpless, and fledge after six weeks in the nest. A pair of lovebirds usually manages to raise three or four offspring a year.

▼ **THE LOOK OF LOVE** Pair of Fischer's lovebirds (*right*) mating. *Far right*: the peach-faced lovebird.

LABOR OF LOVE

1 Caress...
To reaffirm the bond between them, a male peach-faced lovebird treats his mate to gentle preening at the pair's chosen nest site.

3 Load...
Using her bill, she tucks the bark into her soft rump feathers and stows away more strips until her plumage can carry no more.

2 Collect...
Encouraged by his attention, the female flies to a nearby tree and tears off thin strips of bark to use as nest material.

4 Carry
She drops some bark pieces as she flies back to her mate, but holds onto enough to start building her nest.

BEHAVIOR

Like many members of the parrot family, lovebirds are highly sociable. From sunset until daybreak, they roost together. During the day, they're usually seen in flocks of 20 or so, but gatherings of up to 300 occur at a particularly rich food source, such as a fig-laden tree or a field of ripe grain.

When they're not searching for food, lovebirds devote much of their time to their bill and feathers. They repeatedly nibble at hard objects, such as branches, to keep their continuously growing bill at a perfect length. The lovebird uses its feet to take oil from a gland near the base of its tail, then applies it through its plumage. Lovebirds bathe regularly; the gray-headed lovebird cleans its plumage by hanging upside down from a branch in the rain.

Lovebirds communicate with high-pitched calls and by adopting specific poses, such as lowering their head and opening their wings to show hidden colors.

Profile: Lovebirds

Dexterous feet and miniature, lightweight bodies make all lovebirds more agile than many of their larger relatives in the parrot family.

Flight Profile
The combination of a large head, stocky body and short tail with long, tapering wings gives lovebirds a distinctive silhouette in the air.

Bill
Broad, powerful bills crack seeds and tear into fruit. The strongly hooked upper mandible fits into the smaller, lower mandible like a jigsaw piece.

Eyes
Lovebirds' high-set eyes are positioned to give the birds the all-round vision they need to spot danger.

Species shown: red-faced lovebird (*Agapornis pullaria*)

Plumage
A male red-faced lovebird (*right*) wears a mainly green plumage that contrasts sharply with his orange-red bill, cheeks and forehead — which are noticeably more orange in the female. Both sexes have a bright blue patch, or rump, at the base of their tail.

Feet
Two of the short, strong toes point forward and two backward to give a lovebird a firm hold on perches. This versatile arrangement also allows a lovebird to manipulate food with consummate skill and hang upside down without losing its grip.

Creature Comparisons

Identifying lovebirds in the wild can be a challenge, especially where a number of species occur. These six examples illustrate differences among nine species. Masked, black-cheeked, Fischer's and Nyasa lovebirds have broad, naked eye rings; both sexes have similar plumage. Red-faced, Abyssinian and gray-headed (or Madagascar) lovebirds have less prominent feathered eye rings; the sexes have dissimilar plumage. Peach-faced and black-collared lovebirds have characteristics from both groups.

Gray-headed (male) | Black-cheeked | Masked | Fischer's | Peach-faced | Abyssinian (male)

Vital Statistics

Weight	1–2 oz.
Length	5–6.5"
Wingspan	9–12"
Sexual Maturity	1 year
Breeding Season	Varies between species and regions
Number of Eggs	3–8; usually 3 or 4
Incubation Period	22–23 days
Fledging Period	38–50 days; 43–44 days in most species
Breeding Interval	1 year
Typical Diet	Seeds, flowers, fruit, leaf buds; some insects
Lifespan	About 12 years in captivity

Related Species

● Lovebirds, cockatoos, parrots, lories, parakeets, parrotlets and macaws all belong to the *Psittacidae* family, which has about 350 species, including the Senegal parrot, *Poicephalus senegalus* (below).

584 Lovebirds

Luzon Bleeding-heart

- **ORDER** · *Columbiformes*
- **FAMILY** · *Columbidae*
- **GENUS & SPECIES** · *Gallicolumba luzonica*

KEY FEATURES

- Most striking of all the bleeding-heart pigeons; its breast markings resemble a deep, bleeding wound

- Only bleeding-heart pigeon to lay two eggs; all others lay just one

- Spends most of its time on the ground, eating whatever edibles it can swallow whole

- Produces crop milk to feed nestlings

WHERE IN THE WORLD?

Found on only Luzon and Polillo, two islands in the Republic of the Philippines, an archipelago made up of over 7,000 islands; these are part of the Malay archipelago off the southeast coast of Asia

LIFECYCLE

True to its plumage, the Luzon bleeding-heart fulfills the role of the faithful lover — male and female mate for life. But it also plays the victim, crouching when alarmed.

HABITAT

The Luzon bleeding-heart is named for one of the two islands it inhabits. Luzon is the largest of over 7,000 islands in the Republic of the Philippines. The bird also lives on Polillo, a much smaller island to the east. The Philippine Islands are part of the Malay archipelago, the largest group of islands in the world. About 35% of the islands are woodlands, as are Luzon and Polillo. The mountainous regions and fertile plains on these islands offer an abundance of food for the resident bleeding-heart.

The largest freshwater lake in Southeast Asia, Laguna De Bay, is found on Luzon; it is a vital source of freshwater for the Luzon bleeding-heart.

▼ **VOLCANIC LUZON**
The Luzon bleeding-heart is found on the island of Luzon in the Philippines.

 **DID YOU KNOW?**

● Crop milk contains 65–81% water, so parents must drink more water each day to compensate for the drain on their reserves.

● When the Luzon bleeding-heart gets something in its eye, it will rub its eye on its shoulder — this is unusual, since most birds only blink.

BREEDING

The Luzon bleeding-heart mates for life. The male utters a mournful *crooo* to attract a mate. With a second, similar call, the male declares his loyalty. The two calls are distinct, varying in loudness and intensity, in order to communicate the correct message. After bowing and other courtship displays, the birds use small twigs, roots and grasses to build a nest, low in a bush or tree. The female lays two eggs and incubates for about 17 days. During incubation, both the male's and female's crops, an extension of the esophagus, begin to secrete milk, which will be used to feed the down-covered nestlings. After the first few days, the parents add seeds and insects and regurgitate the mixture for the chicks. Luzon bleeding-heart chicks grow quickly with the help of the protein, fat, minerals and vitamins A and B that crop milk supplies. The chicks begin searching for their own seeds and insects by 4 weeks of age.

THE MATING GAME

❶ **Chasing 'round...**
A Luzon bleeding-heart male is strongly attracted to a female and runs after her in an attempt to get her attention.

❷ **Puffing up...**
Interested, the female stops and the courtship show begins. The male inflates his breast enlarging the blood-red patch.

FOOD & FEEDING

The Luzon bleeding-heart is a casual forager — it walks along and turns over soil with flicks of its bill. The bleeding-heart eats fallen seeds, berries, snails, ticks and other insects — almost anything that it can swallow whole, since the bird cannot bite, chew or de-husk its food. If food is not conveniently found on the ground, the bird grabs and tugs to pull suitably sized morsels from plants. The bleeding-heart pigeons all tend to eat more animal food than most other pigeons.

To wash all of the various food items down, the Luzon bleeding-heart drinks at least once a day, usually more. Like all pigeons, the bleeding-heart drinks quickly by sucking in a continuous motion, not lifting the head up to swallow. This ability is unique among birds.

◀ **OUT TO LUNCH**
A bleeding-heart searches the ground for food.

BEHAVIOR

The Luzon bleeding-heart spends most of its time on the forest floor. It searches for food by turning over leaf litter and dirt; it also stretches quite often, for comfort and possibly to keep its wings loose for a quick getaway if predators approach. To stretch, the bird pulls its folded wings upward while lowering its head and tail, and also stretches one leg backward while extending the wing on the same side. When the bleeding-heart does fly, it is usually with its mate to a nesting site or in small flocks to find water. Highly territorial, the male defends its area, first with warning calls and, if necessary, by fighting to the death. The Luzon bleeding-heart is successfully raised in aviaries around the world, but cannot be mixed with smaller species because of its aggressive nature. When confronted with a larger bird of prey in the wild, the bird emits a grunting, panting or gasping sound. This distress call is often sounded at the sight of other predators, including humans. The bird then flies a short distance, lands and continues its escape by running.

▼ **A TOPNOTCH WATCH**
A Luzon bleeding-heart perches and scans for unwelcome intruders.

❸ Bowing down…
The male then lowers his head, with his tail pressed firmly to the ground, and emits a rapid "*croo-coo*."

❹ All grown-up
The female accepts the male's advances and the pair mates; they will stay together for life.

CONSERVATION

In its natural range, the Luzon bleeding-heart is at a low risk of endangerment. The bleeding-heart's limited range and forest habitat make the bird's long-term prospects less positive, since the logging industry destroys millions of cubic feet of timber in the Philippines each year.

Luzon Bleeding-Heart 587

PROFILE LUZON BLEEDING-HEART

The Luzon bleeding-heart appears mortally wounded with its indented breast feathers that are colored blood red.

IN FLIGHT
A strong flier, the bleeding-heart has light-gray wings with dark-red bars; the body's dark-gray feathers have iridescent fringes, giving the bird an amethyst purple or bronze-green appearance in the sunlight.

BREAST FEATHERS
The feathers in the middle of the breast are shorter than the surrounding white feathers. The indentation, with its blood-red color, gives the illusion of a deep gash.

FEET
The dark red feet seem stained from its "wound." The three toes pointing forward and one toe pointing backward allow the bird to perch at its roosts, and also allow it to run during courtship displays.

FEMALE
Eye color is an important distinguishing feature between males and females, since both have similar plumage. The female's iris is purplish-gray, while the male's is blue.

BODY SHAPE
Like all members of the pigeon family, the bleeding-heart has a plump body, a short neck and a small head.

VITAL STATISTICS

WEIGHT	Unknown
LENGTH	12"
WINGSPAN	About 14"
SEXUAL MATURITY	18 months
BREEDING SEASON	March–June
NUMBER OF EGGS	2
INCUBATION PERIOD	About 17 days
FLEDGING PERIOD	About 12 days
BREEDING INTERVAL	Up to 1 year
TYPICAL DIET	Seeds, fruits and invertebrates, including insects
LIFESPAN	Over 20 years

RELATED SPECIES

● The family *Columbidae* includes over 300 species of pigeons and doves in 42 genera, found worldwide except Antarctica. The Luzon bleeding-heart is one of 19 species in the genus *Gallicolumba*. The bleeding-heart pigeons, the less ornamented golden-heart, *G. rufigula*, and the Celebes quail dove, *G. tristigmata*, are found in the Malay archipelago. Most islands contain only one species; New Guinea has three.

CREATURE COMPARISONS

Measuring almost 16" long, the Nicobar pigeon (*Caloenas nicobarica*) is larger than the Luzon bleeding-heart. Both birds live in the Philippines, where they search for seeds, fruit and insects on the ground. But whereas the Luzon bleeding-heart inhabits the Philippines' largest island, Luzon, the Nicobar pigeon is found only on small wooded islands and the islets off the larger land masses. The Nicobar pigeon is also found west of the Philippines on the Nicobar Islands, hence its common name.

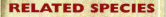

Nicobar pigeon

Luzon bleeding-heart

Macaroni Penguin

• ORDER •	• FAMILY •	• GENUS & SPECIES •
Sphenisciformes	*Spheniscidae*	*Eudyptes chrysolophus*

KEY FEATURES

● Named for its long golden crests, which resemble the flashy feathers worn by 18th-century Englishmen

● Breeds in extremely dense colonies and interacts socially within them

● Monogomous and devoted parents, macaroni penguins will spend up to 45 days fasting when caring for chicks

WHERE IN THE WORLD?

Found on sub-Antarctic islands in the South Atlantic and Indian Oceans and the Falkland Islands off of Argentina; also in the waters south of South America and Africa

LIFECYCLE

Macaroni penguins spend two years with their chick. While one parent ventures out, sometimes for several days, the other parent cares for the young and fasts.

CONSERVATION

With over 11,000,000 breeding pairs, macaroni penguin populations are stable with no major threats, although some species of gull will feed on abandoned eggs, and leopard and Atlantic fur seals will occasionally feed on adults at sea.

HABITAT

The macaroni penguin has two separate habitats. Outside the breeding season, the bird is believed to be completely *pelagic*, living in the open waters of the Antarctic and sub-Antarctic; it cannot be found at nesting colonies or other land areas. Breeding colonies are established on steep rocky slopes, headlands and on level ground areas of the islands falling within its general habitat range. Although the macaroni prefers areas devoid of vegetation, some have been known to nest on patches of tussock grass on the edges of large breeding colonies.

▶ **PLENTY OF PENGUINS**
Macaroni penguin breeding colonies can consist of up to 100,000 breeding pairs.

BEHAVIOR

With such a crowded, complex and social living structure, it is not surprising that macaroni penguins exhibit a wide variety of gestures and vocalizations. Calls seem to be associated with location of a mate, fighting and sexual activity, and tend to vary in length, pitch and associated gestures. Macaronies are extremely vocal during territory establishment and pair formation, with much chattering and trumpeting. However, despite their often massive size, breeding colonies tend to be rather quiet during incubation, with occasional spurts of activity occurring whenever one parent returns from a foraging trip and during the "changing-of-the-guard" at the nest. Parents recognize each other as well as their chicks by voice; they use loud trumpeting from a distance and a quivering, chattering call when close up. Quite social with each other, some colonies will also nest in harmony with different species of albatross living just outside their colony.

▲ **TAKING THE PLUNGE**
A penguin prepares to dive in for a hunt.

▲ **ROCKY HOME**
Macaroni penguins prefer steep, rocky slopes.

LIGHTNING STRIKE

① Do not disturb…
Macaroni penguins attempt to make their way through the crowded nesting site without disturbing other birds by following access routes.

③ Take that!
Sometimes the aggression may lead to bill jousting, where both parties will lock bills and attempt to force each other to the ground.

BREEDING

Macaroni penguins tend to be at sea from April through October and arrive at colonies between October and November. Once they find their mate within the colony, they greet each other with much trumpeting, head shaking and bowing. The pair, which mates for life, will then construct a nest consisting of a shallow scrape in the mud or gravel lined with a few small stones. In late November to December, a clutch of two eggs is laid, the first being about 60% smaller then the second. Incubation takes about 33–37 days; both parents incubate the eggs for the first 8–12 days, the female incubates them for the following 12–14 days, and the male incubates them for the last 9–11 days. Hatching of the young takes 24–48 hours from pipping (first breaking through the shell) to emergence from the egg, but usually just the larger second chick survives. The father guards the nest for another 25 days while the female forages for food. After 25 days several chicks will convene together to form nurseries. The father then returns to sea to forage. For about 10–20 days the chicks will be fed by both parents at increments of about every 1–2 days, and the fledging period lasts about two months.

▶ **CHANGING SHIFTS**
Both parents take turns protecting the vulnerable nest and greet each other vocally whenever one returns after searching for food for the chick.

FOOD & HUNTING

During the nonbreeding season, the macaroni penguin feeds primarily on crustaceans, cephalopods and small fish. When the chicks are young, they are fed crustaceans, with cephalopods and small fish being added to the diet as the chicks grow larger. The tongue and palate are equipped with spines in order to better grip fish, squid, krill and other slippery prey. The macaroni usually forages during the day, with trips averaging 12 hours, but during chick-rearing, trips last 25–50 hours. Macaroni penguins are superb divers; during a long foraging trip, they spend about 10% of their hunting time diving underwater in search of prey. When diving, they can reach depths of almost 400', although they average about 130' during the day and 10' at night. Though they can be underwater for several minutes, the average length of a dive is 1.5 minutes.

❷ No trespassing…
Although usually peaceful, other birds, when disturbed, may exhibit aggressive behavior with loud calls and even flipper smacks.

DID YOU KNOW?

● Some males have been known to make foraging trips of up to 270 hours.

● The macaroni penguins were named after a group of Englishmen who wore fancy feathers in their hats and introduced Italian macaroni to England.

▼ **MACARONI MENU**
Macaroni penguins take to the water to hunt fish.

❹ Home sweet home
When the penguin reaches home, the pair greet each other with vocalizations and bowing as they prepare for nest relief.

PROFILE: MACARONI PENGUIN

With its three layers of feathers and stores of fat, the macaroni penguin is superbly equipped for living in the chilly waters of the Antarctic.

FACE
The black face of the macaroni penguin distinguishes it in appearance from its close relative, the royal penguin.

BILL
The large, bulbous bill has ridges at the base of the upper mandible to give it a better grip on fish, squid or other slippery prey.

JUVENILE
At 2 years old, the macaroni has acquired short plumes and is ready to molt and leave the colony.

CHICK
Young chicks lack the distinctive head plumes, which do not begin to appear until 1 year of age.

FLIPPER
The strong, stiff flippers, which lack flight feathers, can propel the macaroni through the water at speeds of 6 mph.

FEET
The webbed feet are pink on top with black soles that help control heat loss; when swimming, the penguin's feet serve as rudders in conjunction with the tail.

VITAL STATISTICS

WEIGHT	11–13 lbs.
HEIGHT	28"
SEXUAL MATURITY	Female 5 years; male 6 years
BREEDING SEASON	October–December
NUMBER OF EGGS	2, with only 1 surviving
INCUBATION PERIOD	60–70 days
FLEDGING PERIOD	2 years
TYPICAL DIET	Fish, squid, crustaceans, and krill
LIFESPAN	Unknown

CREATURE COMPARISONS

The royal penguin (*Eudyptes schlegeli*) is nearly identical to the macaroni penguin; the main difference is the white face of the royal relative. The royal penguin is slightly larger than the macaroni penguin, but these two species are so similar that many scientists regard the royal penguin as a subspecies of, or merely a color variation of, the macaroni penguin. Some macaroni penguins have appeared with coloring similar to the royal penguin's, but it is unknown whether these are mutations or hybrids. The royal penguin consists of about 85,000 breeding pairs in about 57 colonies all located on Macquarie Island off of southern Australia.

Royal penguin

Macaroni penguin

RELATED SPECIES

● The *Spheniscidae* family consists of 6 genera and 17 species of penguins, all of which are flightless birds living in or around Antarctica. The genus *Eudyptes* is the largest of the genera and is made up of 6 species of crested penguins. All penguins stand upright and walk with a shuffling gait. They range in size from the little blue penguin, *E. minor*, to the emperor penguin, *Aptenodytes forsteri*.

592 Macaroni Penguin

Magnificent Frigatebird

• ORDER •
Pelecaniformes

• FAMILY •
Fregatidae

• GENUS & SPECIES •
Fregata magnificens

KEY FEATURES

- An aggressive pirate of the skies, it forces other seabirds to give up their catch of fish
- Long wings, a streamlined shape and low weight give it great maneuverability in the air
- Snatches fish from the sea's surface and steals eggs and chicks from other birds' nests

WHERE IN THE WORLD?

Along coasts from Florida to Brazil and Baja California to Ecuador; on the Galapagos Islands and off the West African coast on the Cape Verde Islands

Magnificent Frigatebird 593

 LIFECYCLE

Despite being a poor swimmer and lacking waterproof feathers, the magnificent frigatebird is an ocean predator, notorious as a fearless pirate of the tropical seas.

HABITAT

▲ **LOCAL BRANCH** Mangroves are popular roosting and nesting sites.

The frigatebird is found along the tropical coasts of the Americas and western Africa. Warm trade winds, which blow year-round with variable force toward the equator, have an influence on its distribution. These winds produce thermals night and day, letting the frigatebird soar freely while searching for squid, a favorite food, which come close to the sea's surface at night.

The frigatebird is closely tied to mainland coastal areas and is rarely seen mid ocean. It breeds along coasts or on small offshore islands, especially in stands of mangrove trees. On the Galapagos Islands, it may even nest in a large cactus plant. Where vegetation is scarce, such as on the Cape Verde Islands, off West Africa, the frigatebird may be forced to construct a nest on the bare ground.

 **DID YOU KNOW?**

- "Frigatebird" may come from the comparison of the bird to the fast frigate ships once used by pirates to attack merchant vessels.
- The magnificent frigatebird has the highest wingspan-to-weight ratio of all seabirds. The skeleton is less than 5% of total weight.

CONSERVATION

With a world population of several hundred thousand, the magnificent frigatebird isn't under threat, but habitat destruction is a potential danger. The tiny colony on the Cape Verde Islands has been reduced to about 12 pairs. Precise figures are difficult to assess because some non-breeding birds remain at sea in the nesting season.

BEHAVIOR

▲ **FISH OUT OF WATER** A juvenile swallows its catch, head first, in flight.

FOOD & FEEDING

The frigatebird follows schools of dolphins or tuna and uses its agility to snatch fish that break surface to escape underwater predators. Squid and jellyfish are also part of its diet. These are usually taken in a more leisurely fashion; the frigatebird drifts down from a height to grab its prey from the surface of the sea.

Seabirds' eggs and chicks, particularly those of terns, are plundered from nesting colonies, and hatchling turtles are scooped up from the beaches. Magnificent frigatebirds have, like many other seabirds, learned to exploit the opportunities presented by humans and will follow fishing boats to feed on the scraps and offal thrown overboard.

The frigatebird's reputation rests on its spectacular pursuit of other seabirds, harassing them until they give up their catch.

BOOBY PRIZE

❶ Unsuspecting... A red-footed booby returns from a successful fishing trip, as yet unaware of the aerial pirate about to swoop to the attack.

❷ Pirate ahoy... Using its superior flying skills, the frigatebird homes in on its victim, tugging fiercely at its wing and tail feathers with its sharp bill.

Leaving its roost on land in early morning, the frigatebird soars out to sea in search of food. It forages alone, but may congregate around concentrated sources of food, such as fishing boats. When not soaring, it rests on perches provided by ships' masts, buoys and fishing posts.

Dissipating excess heat is a problem for birds in tropical climates, but the frigatebird prevents overheating in a couple of ways. Males and females use 'gularfluttering.' This is similar to panting and involves the bird passing air over the mucous membranes in its throat. Mucus then evaporates, causing heat loss. Also, by ruffling its feathers, the bird lets the breeze draw heat directly away from its skin.

Unlike most other seabirds, the frigatebird drinks freshwater when it has the opportunity, flying low over the surface and scooping water into its bill.

❸ Boarding party…
In danger of injury through being knocked out of the sky, the harassed booby regurgitates part or all of its fish load and quickly flies off.

❹ Booty
As the prize falls rapidly to the sea, the frigatebird dives after it, swooping down and deftly catching the fish in midair.

BREEDING

▲ **HOME TO ROOST**
When this male attracts a mate, the female will organize the nest and add material of her own.

The cue for the frigatebird to start breeding arrives when the trade winds begin to blow strongly. Then, the male collects twigs, leaves and seaweed for the female who constructs a flimsy nest platform, usually 7–17' up in a tree. Frigatebirds often nest close to each other to form a colony. Both parents incubate the egg and feed the chick when it hatches. Adults may have difficulty catching enough food to feed their young and often resort to stealing it from other birds.

The frigatebird has one of the longest breeding cycles of all seabirds because of the chick's slow development and the long period of care (5–7 months) needed after leaving the nest. The female breeds only once every two years.

▼ **FEATHER WEIGHT**
The single chick is naked on hatching, but soon develops fluffy white down.

Magnificent Frigatebird 595

PROFILE MAGNIFICENT FRIGATEBIRD

Well adapted to an aerial existence, the magnificent frigatebird is a marauder of the tropical coasts of North and South America.

HOOKED BILL
The bill is long and pointed and has a sharply hooked tip, letting the frigatebird grasp slippery prey, such as flying fish and squid. When harassing other birds, the bill is also used as a menacing weapon.

THROAT SAC
The male inflates his bright red throat pouch to attract females. The bill is vibrated against the sac to produce a peculiar drumming sound.

FEMALE
The sexes look quite different. Larger than the male, the female lacks the inflatable throat pouch and has a large white area on her breast that stands out against the black plumage.

FEET & CLAWS
The frigatebird's toes are webbed. However, as the bird rarely needs to swim, the webs are very small. The feet are equipped with strong claws to aid perching in nesting trees.

VITAL STATISTICS

WEIGHT	2–3 lbs.
LENGTH	3–4'
WINGSPAN	7–8'
BREEDING SEASON	Throughout the year; in some locations favors dry season
NUMBER OF EGGS	1
INCUBATION PERIOD	40–50 days
FLEDGLING PERIOD	20–24 weeks
BREEDING INTERVAL	2 years
TYPICAL DIET	Flying fish, squid, offal, scraps, seabird eggs and chicks
LIFESPAN	Up to 30 years; average 12–15

RELATED SPECIES

- There are five species of frigatebird in the genus *Fregata*; great frigatebird (*Fregata minor*) has the widest distribution. *F. aquila* is confined to breeding on Ascension Island and *F. andrewsi* to Christmas Island. The smallest is the lesser frigatebird, *F. ariel*. Frigatebirds belong to the *Pelecaniformes* order, which also includes pelicans, gannets, boobies and cormorants.

CREATURE COMPARISONS

The red-footed booby (*Sula sula*) has the same range as the magnificent frigatebird and often nests in the same colony. The frigatebird is very buoyant in the air; the booby appears cumbersome, with its heavy body and labored flight. When feeding, the booby dives vertically from 30–100' into the sea, submerging itself for fish and propelling itself by its webbed feet. Unlike the frigatebird, which rarely gets wet, the booby has well-developed oil glands above the tail that provide essential waterproofing.

Magnificent frigatebird

Red-footed booby

Mallard

· ORDER ·
Anseriformes

· FAMILY ·
Anatidae

· GENUS & SPECIES ·
Anas platyrhynchos

KEY FEATURES

- The world's most abundant species of duck
- Feeds on submerged plants by upending and pushing its long neck underwater
- Male abandons his mate after egg-laying
- Ducklings stay close to their mother for protection, often swimming behind her in a long line

WHERE IN THE WORLD?

Found in a huge range that includes nearly all of Northern Hemisphere's temperate zone; introduced populations in Australia and New Zealand

LIFECYCLE

The mallard can raise a family in any wetland habitat, including the smallest pool and park lake, but each pair must produce a large brood to ensure enough chicks survive.

HABITAT

The mallard prefers to live in shallow waters with lush bankside vegetation. Although it can adapt to most every type of wetland, it ignores the deepest lakes and open seas. The mallard rarely moves far from its breeding grounds, but some fly to estuaries in winter, where there is a rich supply of food all year-round.

Mallards are tolerant of humans and are a common sight in garden ponds, park lakes and sewage farms.

▲ **PARK LIFE**
Even busy parks have quiet corners for nesting.

CONSERVATION

Although hunted, the mallard has a stable world population of 28 million birds. However, in some areas the species' genetic purity is threatened by breeding with domesticated ducks.

FOOD & FEEDING

The mallard employs a number of methods to find food. Out on the water, it paddles with its feet to disturb food items, which it then snaps up from the surface, or it sieves water through the sides of its bill to extract tiny animals. It up-ends in the water to browse water weed, or grab seeds, small snails and invertebrates from the bottom. On land, the mallard grazes short grass, browses shoots, and picks up whatever food it comes across, including scraps thrown by humans.

▲ **FILTER FEEDERS**
Souplike mud at the water's edge teems with a wide range of invertebrate prey.

BEHAVIOR

Outside the breeding season, mallards gather in flocks of hundreds or even thousands of birds, which tend to take part in the same activity at the same time. For example, the whole flock may feed, preen, bathe or simply rest together. In late winter, the flocks disperse into groups of several pairs each to search for suitable nest sites. But, even during the breeding season, neighboring pairs of mallards often associate in loose groups.

# BREEDING

DID YOU KNOW?

- Only female mallards can "quack." The male has a feeble, high-pitched call.

- Both sexes molt their feathers in midsummer and become flightless for about 4 weeks.

- Pools heated by thermal springs allow the mallard to stay in Iceland all winter.

- Young mallards have many enemies, including bass, snapping turtles and raccoons.

Groups of male mallards chase and display to females all winter long, but especially in February and March. Once a drake has attracted a mate, he stays at her side to prevent her mating with other males. When she has laid her clutch of 9–13 eggs, he takes no part in rearing the young. The downy chicks leave the nest 24 hours after hatching, and can feed themselves, but need their mother's protection for two months. Even so, as few as one in ten reach adulthood.

▲▶ SINGLE PARENT
The drake (*right*) deserts his mate, so she cares for her large family alone.

◀ WELL GROOMED
Preening and wing-stretching maintain feathers.

▲ FEEDING FRENZY
Large flocks soon form at rich feeding grounds.

WATER OFF A DUCK'S BACK

❶ Land...
Several mallards drop down to a lake to feed. They spread their feet to act as brakes; just before hitting the water, they flap backward.

❷ Feed...
While the mallards are upending to browse underwater weed, a peregrine falcon flies overhead in search of its own meal.

❸ Fly...
Most of the mallards spot the falcon and leap into the air to fly to safety, but one drake is unaware of the danger and continues to feed.

❹ Dive
The lone mallard is, in fact, safe on the water, as the falcon hunts only in midair. The hunter heads off in search of airborne prey.

Mallard 599

Profile Mallard

A highly versatile body plan allows the mallard to exploit a wide range of feeding opportunities, making it the world's most successful duck.

FEMALE

The female, or duck, is slightly smaller than the male, or drake, and has a drab, brown-streaked plumage to camouflage her when sitting on the nest.

BILL
The sharp "nail" on the tip of the bill is used to graze grass and pick up small prey. The mallard can also filter tiny animals from the water by means of the comblike membranes, known as *lamellae*, that line the inside of its bill.

FEET
Webbed feet enable the mallard to paddle efficiently and half-submerge itself to reach aquatic plants.

SPECULUM

Male and female alike have a bright blue patch, known as a *speculum*, on each wing. The specula are much more visible in flight.

VITAL STATISTICS

WEIGHT	1.5–3.5 lbs.
LENGTH	20–26"
WINGSPAN	30–40"
SEXUAL MATURITY	1 year
BREEDING SEASON	February–June
NUMBER OF EGGS	9–13
INCUBATION PERIOD	About 28 days
FLEDGING PERIOD	50–60 days
BREEDING INTERVAL	1 year
TYPICAL DIET	Shoots and seeds of aquatic plants, grass, insects, snails, worms
LIFESPAN	Up to 29 years in captivity

RELATED SPECIES

● The mallard is 1 of 36 species of surface-feeding duck in the genus *Anas*, which also includes the Eurasian wigeon, *A. penelope* (male, below), and the pintails, shovelers and teals. Ducks are in the family *Anatidae* in the order *Anseriformes*, which includes geese and swans.

CREATURE COMPARISONS

From a distance, the northern shelduck (*Tadorna tadorna*) seems to have a patterned black-and-white plumage, as the species' green head and orange-chestnut breast band are visible only up close. The male shelduck has a prominent "knob" on the bill.

A bird of Eurasia's coastal mudflats and estuaries, the shelduck is rarely found far inland. It has a more specialized diet than the mallard, feeding on small shellfish, snails and worms from the mud.

Mallard (male)

Northern shelduck (male)

600 Mallard

Malleefowl

• ORDER •
Galliformes

• FAMILY •
Megapodiidae

• GENUS & SPECIES •
Leipoa ocellata

KEY FEATURES
- Builds large mounds of soil and plant matter in which it lays its eggs
- Eggs are incubated using heat generated by rotting matter inside the mound and by the sun's rays
- Rare across its range; the IUCN (World Conservation Union) classifies it as vulnerable

WHERE IN THE WORLD?

Once distributed over large areas of southern Australia; now restricted to scattered semiarid areas of the southwestern and southern mainland

LIFECYCLE

The hard-working malleefowl faces a life of unremitting toil. Although only a small bird, it builds a huge nesting mound and spends up to 11 months of the year tending it.

HABITAT

▲ **MOVING MOUNTAINS**
The mound may be up to 17' in diameter and 5' high.

The malleefowl needs a reliable food supply of seeds and fresh young shoots. It finds this in the vast expanse of semiarid land in southern Australia. This region encompasses areas where mallee (a dwarf eucalyptus) and acacia are abundant.

Preferring open areas at ground level, the malleefowl looks for territory with shrubby undergrowth nearby in which to find food, and an almost complete canopy above to shield it from the sun.

 **DID YOU KNOW?**

- If an adult malleefowl comes across an emerging chick when tending the mound, it simply kicks the chick out as if it were an item of mound material.

- The malleefowl may move up to 300 tons of soil a year regulating the mound's temperature; 35 cu. feet may be added or removed at a time.

BREEDING

The malleefowl uses heat, generated by sun and decomposition of buried vegetation within its mound, to incubate its eggs. The male begins building the mound in early winter, digging a hole 10–13' in diameter and 3' deep, then filling it with vegetation.

In early spring, once rain wets the vegetable matter and begins the decomposition process, the birds dig a chamber into which the female lays her eggs. The parent birds then cover the whole mound with sandy soil. Throughout incubation the birds add or remove soil to maintain an ideal temperature of 91°F.

Upon hatching, each chick digs to the surface where it must fend for itself. It can fly almost immediately.

▶ **BIRTHDAY SUIT**
The hatchling already has camouflaging plumage.

TENDING THE NEST

① Thermometer bird...
Using heat-sensitive areas in its mouth, the malleefowl tests the mound's temperature, which should be about 91°F.

② When the going gets hot...
If the mound feels too warm, the malleefowl scrapes away some of the soil and vegetation to let the excess heat escape.

FOOD & FEEDING

602 Malleefowl

BEHAVIOR

Although the malleefowl mates for life, the male and female are solitary and are seldom seen together away from the mound. The male tends the mound every hour, and so forages, sunbathes and roosts close to it. The female, however, wanders across the pair's range in search of food to sustain her heavy egg production; she visits only for short periods.

The malleefowl's range varies from 0.4–1.6 sq. miles. It is smaller in rainy areas due to greater densities of malleefowl.

▼ **TOP OF THE HEAP**
Old mounds are often reused in later years.

3 Chilled out…
The malleefowl replaces mound material once the temperature reaches the correct level. The male does the majority of the digging.

4 Breakthrough
After hatching, a chick digs through 3' of soil to reach the surface, a feat that leaves it exhausted for several hours.

The adult malleefowl eats a variety of plant matter including seeds, buds, flowers, leaves and fruit. It typically searches for food by walking slowly along the woodland floor, scratching among leaf litter and in the soil with its powerful feet. Without actively seeking them, the malleefowl also eats any invertebrates and insects that it finds, inadvertently picking them up while foraging.

The young chick feeds almost entirely on invertebrates at first; they are readily available and more easily converted into energy and muscle than plant matter. It gives the chick the best possible start in life — which is vital since it has to fend for itself as soon as it leaves the mound. As it matures, the young malleefowl eats a greater proportion of plant matter until, as an adult, its diet consists mostly of tender young shoots and seeds.

◀ **SURROUNDED BY FOOD**
The malleefowl often eats sand to grind down food.

CONSERVATION

The total population of malleefowl is now thought to be fewer than 10,000, following a dramatic decline in recent years. The greatest threat to the species is the destruction of its habitat by wheat-growing and sheep-farming. Many eggs are taken by the introduced European fox and there is significant predation by feral cats.

Profile: Malleefowl

Specially adapted legs and feet enable the chicken-sized malleefowl to move huge amounts of soil and vegetation to construct nesting mounds.

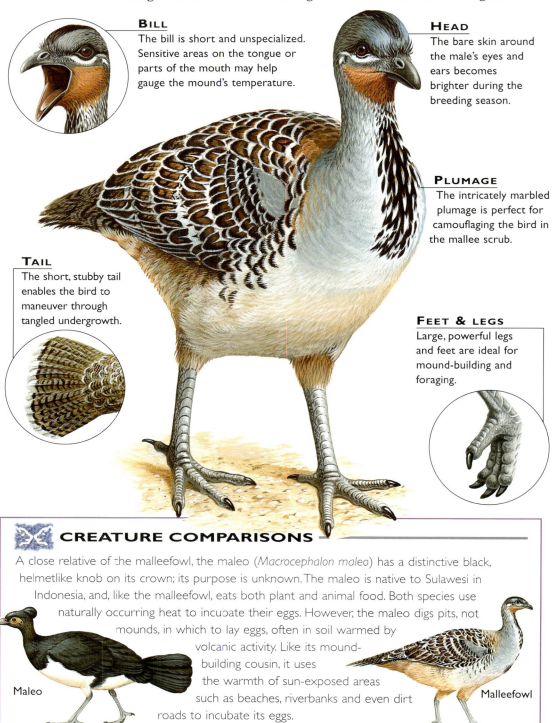

BILL
The bill is short and unspecialized. Sensitive areas on the tongue or parts of the mouth may help gauge the mound's temperature.

HEAD
The bare skin around the male's eyes and ears becomes brighter during the breeding season.

PLUMAGE
The intricately marbled plumage is perfect for camouflaging the bird in the mallee scrub.

TAIL
The short, stubby tail enables the bird to maneuver through tangled undergrowth.

FEET & LEGS
Large, powerful legs and feet are ideal for mound-building and foraging.

CREATURE COMPARISONS

A close relative of the malleefowl, the maleo (*Macrocephalon maleo*) has a distinctive black, helmetlike knob on its crown; its purpose is unknown. The maleo is native to Sulawesi in Indonesia, and, like the malleefowl, eats both plant and animal food. Both species use naturally occurring heat to incubate their eggs. However, the maleo digs pits, not mounds, in which to lay eggs, often in soil warmed by volcanic activity. Like its mound-building cousin, it uses the warmth of sun-exposed areas such as beaches, riverbanks and even dirt roads to incubate its eggs.

Maleo | Malleefowl

VITAL STATISTICS

WEIGHT	Male 5–6 lbs.; female 3–5 lbs.
LENGTH	24"
WINGSPAN	12–13.5"
SEXUAL MATURITY	Usually 4 years
BREEDING SEASON	June–February
NUMBER OF EGGS	15–24
INCUBATION PERIOD	62–64 days average
FLEDGING PERIOD	Independent on hatching
BREEDING INTERVAL	Annual but not during droughts or rainy periods
TYPICAL DIET	Omnivorous, but mainly eats buds, flowers, fruits, seeds
LIFESPAN	Up to 25 years

RELATED SPECIES

● The malleefowl belongs to genus *Leipoa*, family *Megapodiidae*, a group of birds found almost exclusively in the Australasian region. There are 7 genera and 19 species within the family, the largest genus being *Megapodius*. The *Megapodiidae* are among the most distinctive members of order *Galliformes*, which contains 274 species including pheasants and curassows.

Mandarin Duck

- ORDER -
Anseriformes

- FAMILY -
Anatidae

- GENUS & SPECIES -
Aix galericulata

KEY FEATURES

- Extravagant plumage of the drake (male) makes it one of the most easily recognized of all ducks
- Native to Asia, but escapees from waterfowl collections established populations in Britain
- Prefers wooded habitats near water and usually nests in tree holes

WHERE IN THE WORLD?

Occurs naturally in eastern China, eastern Russia and Japan; feral populations exist in parts of Europe — the majority in southeast England

LIFECYCLE

Though it seldom strays far from rivers and lakes, the mandarin duck is just as much a woodland bird as a waterfowl, adapted for flying, perching and nesting among trees.

HABITAT

The mandarin duck's natural haunts lie in broad-leaved, temperate forests of northeast Asia. It typically lives along forest streams or pools lined with thick bankside vegetation (trees, shrubs or reeds). Open ground and broad expanses of water are usually avoided, but small flocks do enter rice fields to feed after breeding. Birds from the north of the range migrate south each year to spend the winter in the milder climate south of China's Yangtze River.

As many as 300–400 pairs live in the wild in Britain. Escapees from collections have established a number of breeding areas; colonies exist in most counties of southeast England, as well as sites in Norfolk, Cheshire, Gloucestershire and Tayside.

▼ **IDEAL HOME**
The mandarin duck favors woodland near water.

CONSERVATION

The status of the mandarin duck in its Asian homeland is unclear — few studies have been undertaken there. Because of forest clearance and other habitat disturbances in Asia, conservationists suspect its numbers are declining. However, the wild population in Britain is slowly on the rise.

FOOD & FEEDING

The mandarin duck finds food both in and out of the water. It forages among debris on the bank, dabbles at the water's edge or while swimming and occasionally up-ends to reach deeper, submerged food.

Vegetation makes up the bulk of its diet, but the duck will also snap up small creatures (mostly insects); at certain times of the year, animal food predominates. In early autumn, land snails are an important food source. Flocks studied in Russia eat a variety of seeds (including those from aquatic plants), acorns, insects (such as beetles) and small fish. In Britain, the summer diet consists of waterside and aquatic insects; acorns, beechmast and chestnuts provide sustenance the rest of the year.

▼ **FEELING PECKISH**
Though known as a surface-feeding duck, the mandarin duck feeds in a variety of ways in and out of the water.

606 Mandarin Duck

BEHAVIOR

Outside the breeding season, when pairs have finished tending their nests, mandarin ducks are social birds that gather in flocks (sometimes more than 60). The duck is most active in mornings and evenings, but feeding continues intermittently throughout the day and night. The mandarin duck is perfectly at home on the water or land, both swimming and walking with ease.

The agility of the mandarin duck extends to its power of flight. With strong, rapid wingbeats, it can rise steeply into the air from the water surface or land. This, and its ability to twist and turn tightly in flight, lets the duck negotiate its way at speed through the confined spaces of its woodland home.

▲ **SITTING PRETTY** Perching ducks take their name from their ability to "sit" on slender branches.

BREEDING

Antics of courtship take up a good deal of a mandarin duck's time. Starting as early as September, well before the spring mating season, flocks of drakes gather for communal displays designed to impress females.

By spring, most females will have selected a partner; the pairs engage in further displays to strengthen their bonds prior to mating. During the nesting period, the drake stays nearby while the female incubates the eggs. A few days after hatching, the ducklings feed themselves. In six weeks, they're fully independent.

THE HOLE STORY

❶ Tree house…
Old trees typically provide the preferred nesting cavities. The favorite nest site is a hole several feet above the ground.

❸ Egg factory…
Laying at a rate of one per day, the female produces 9–12 eggs. Once the last is laid, the female begins a four-week stint of incubation.

❷ Home-making…
No nesting material is taken into the hole. The female uses her body to mold a depression in debris already there and lines it with feathers.

❹ Flying the nest
Shortly after hatching, ducklings scramble to the light and launch themselves from the nest entrance under the watchful eye of mother.

 DID YOU KNOW?

● The first captive mandarin ducks were brought to Britain as early as 1747. In 1971, the species was formally accepted on the list of Britain's wild birds.

● A mandarin duck that escaped from London's St. James's Park in 1930 turned up months later in Hungary, 900 miles away.

● In ancient China, the mandarin duck was a symbol of faithfulness, and newlyweds were presented with a pair of live ducks as a good luck token.

Mandarin Duck 607

PROFILE: Mandarin Duck

With rich colors, ornate patterns and fanciful "whiskers" and "sails," the male mandarin duck is one of the most beautiful of all waterfowl.

PLUMAGE
Immediately after the breeding season, the drake molts his colorful feathers and takes on "eclipse" plumage, which is similar to the female's.

FEMALE
Though dowdy in comparison with the drake, the female's prominent eye ring and eye stripe distinguish her from other gray-brown ducks.

WING FEATHERS
The most unusual feature of the drake's plumage is the pair of orange "sails" formed from an enlarged and upturned set of wing feathers.

WINGS & TAIL
Strong wings that beat rapidly and a long tail give the bird excellent maneuverability, compared to other ducks, when flying among trees.

FEET
Sharp claws on the toes of its webbed feet help the mandarin duck get a secure grip when perching on branches.

VITAL STATISTICS

WEIGHT	1–1.5 lbs.
LENGTH	16–20"
WINGSPAN	27–30"
SEXUAL MATURITY	1 year
BREEDING SEASON	Spring
NUMBER OF EGGS	9–12
INCUBATION PERIOD	4 weeks
FLEDGING PERIOD	6 weeks
BREEDING INTERVAL	1 year
TYPICAL DIET	Seeds and nuts; some insects and snails
LIFESPAN	3–6 years

RELATED SPECIES

● The mandarin duck is one of 13 species of perching ducks and perching geese in the *Cairinini* tribe. Different species are scattered across the globe, but all share adaptations to live in wooded habitats. The largest is the spur-winged goose of Africa; others include the muscovy duck of Central and South America, the African pygmy goose and maned wood duck of Australia.

CREATURE COMPARISONS

The wood duck of North America is the mandarin duck's cousin. It, too, is a woodland dweller that nests in tree holes; it is also a popular captive bird in Europe. Though similar in size and shape to the mandarin drake, the male has plumage embellished with glossy greens and browns and bold, white stripes. The female, like the mandarin duck, is mostly gray-brown.

Male wood duck

Male mandarin duck

Manx Shearwater

• **ORDER** •
Procellariiformes

• **FAMILY** •
Procellariidae

• **GENUS & SPECIES** •
Puffinus puffinus

KEY FEATURES

- Spends eight months a year out at sea, following currents and winds for fish and squid
- Glides effortlessly, skimming the ocean waves and only rarely flapping its long, slender wings
- Nests on remote islands, but travels to and from its burrow by night as it's defenseless on land

WHERE IN THE WORLD?

Breeds on offshore islands in the northeastern Atlantic (Iceland south to Madeira), along the east coast of North America and in the Mediterranean; wanders the South Atlantic from August until April

LIFECYCLE

A master of navigation and energy-saving flight, the Manx shearwater spends much of its life flying over the open Atlantic, coming to land for only a few months each year.

HABITAT

▲ **Nest holes**
Colonies look deserted by day: all birds are at sea or incubating underground.

One of the Atlantic Ocean's great wanderers, the Manx shearwater only comes ashore to breed, nesting on offshore islands in huge colonies. It arrives at its Northern Hemisphere breeding grounds in April: the timing depending on the latitude of each colony.

In the breeding season (just over four months), the shearwater stays within a couple hundred miles of its colony. In August, it begins a marathon journey to the South Atlantic, where it spends the Southern Hemisphere's summer in the warm seas between Brazil, Uruguay and Argentina in the west and South Africa in the east.

? DID YOU KNOW?

- The Manx shearwater is named after the Isle of Man, in the Irish Sea, but it no longer breeds there.
- The wailing of breeding Manx shearwaters was once thought to be evil spirits.
- By the time it fledges, a well-fed juvenile can weigh twice as much as its parents.
- A leg-banded juvenile was found in Brazil 17 days after it left its burrow in Britain — over 4,350 miles away.

BEHAVIOR

The shearwater is a highly social bird and is rarely seen on its own. It flies in tightly packed flocks, which dart over the sea like squadrons of low-flying aircraft, twisting and turning in sequence to take advantage of the air rising up between waves. As it rarely has to flap its wings, flying is effortless and uses little energy.

The shearwater's amazing "homing" ability, which isn't fully understood, but may involve an internal "map" of the sun and stars, lets the bird find its way across expanses of open ocean, with unerring accuracy.

▲ **Nocturnal crooner**
Silent most of the year, the shearwater cackles and "moans" at its colonies.

BREEDING

▲ **Finding its way**
A shearwater finds its burrow, remembering certain "landmarks."

▶ **Change of owner**
The shearwater may nest in an old rabbit burrow; the Atlantic puffin may use it next.

610 Manx Shearwater

CONSERVATION

The Manx shearwater faces few threats at sea. It isn't dependent on a single type of prey and is less vulnerable to oil slicks than seabirds that spend most of their time on the water's surface. But the story is mixed at its colonies. On some islands, introduced predators (foxes, cats and rats) reduced local populations. However, where islands have been made into nature reserves, the bird thrives.

FOOD & FEEDING

▼▲ **SUNSET STRIP**
As the sun sinks, the birds fly from their feeding grounds at sea *(above)* **to their colony** *(below)*, **to feed their chicks.**

The shearwater hunts fish, small squid and crustaceans that live close to the sea's surface. It catches its prey in two different ways: by snatching it from the surface while paddling or by plunging underwater to give chase. Its dives are fairly shallow and last just a few seconds, but the shearwater has been found trapped in lobster traps in over 100' of water, proving that it dives deeply if necessary.

Where food is plentiful, hundreds or thousands of shearwaters gather on the sea in scattered groups known as *rafts*; squabbles may break out, but the birds usually feed together in peace. When parents have a chick to feed, they make the nightly flight back to their burrow with food; at other times, they stay at their feeding grounds. Unlike other seabirds, such as gulls and gannets, the Manx shearwater doesn't follow ships to pick up scraps thrown overboard.

In the early spring, shearwaters congregate off coastal islands to prepare for their annual return to dry land. This takes place under the cover of darkness, as their awkwardness on land makes them vulnerable. Birds that have bred in previous years search out their old nest burrows and renovate them, but young adults must find an abandoned burrow or dig one. With thousands of shearwaters breeding close together, the competition for burrows is intense.

The female lays one large egg, which both parents incubate in shifts for 7–8 weeks: while one bird is sitting on the egg, the other feeds at sea, often wandering hundreds of miles away. Once the chick hatches, the male and female leave it unattended during the day while they hunt, each feeding it at night. It quickly becomes fat, but by the end of August, its parents stop feeding it and leave the colony. Driven by hunger, the chick makes the hazardous journey down to the sea and sets off on its long journey south for the winter.

AT THE END OF THE DAY...

❶ Feeding party...
By day a raft of shearwaters may float on the surface, plundering a large shoal of surface-feeding fish.

❷ Homeward bound...
As dusk falls, flocks of birds begin to leave, "pattering" along the sea to gather speed to take off.

❸ Crash landing...
Back at the colony, each shearwater finds its own burrow, flopping down on legs poorly adapted for walking.

❹ Messy business
A parent crawls into the nest to regurgitate a mix of oil and partly digested fish for its chick.

Manx Shearwater 611

Profile   Manx Shearwater

Gliding on its long, straight wings, the Manx shearwater is completely at home above the vast open ocean, where it can smell prey from afar.

BILL
The narrow, but sturdy, bill is hooked at the tip to keep a firm grip on prey snatched from the surface of the sea or caught during short dives beneath the waves.

WINGS
The shearwater flies with rapid wingbeats, then glides on stiff, outstretched wings.

TAIL
Broad and wedge-shaped, the tail helps the bird bank (turn steeply) to gain speed and lift from gusts of wind.

NOSTRILS
The shearwater has a relatively good sense of smell compared to most other birds. Its two external nostrils are joined to form a tube over the top of the bill.

FEET
Webbed feet drive the bird through the water when chasing prey. In flight, they're stowed away under the tail with the webbing closed.

VITAL STATISTICS

Weight	12–16 oz.
Length	1'
Wingspan	3'
Sexual Maturity	5 years
Breeding Season	April–August
Number of Eggs	1
Incubation Period	51–54 days
Fledging Period	70 days
Breeding Interval	1 year
Typical Diet	Fish, small squid and crustaceans
Lifespan	Up to 20 years

CREATURE COMPARISONS

It's difficult to tell shearwaters apart, since they fly fast and low over the ocean, disappearing into wave troughs and only come to land after sunset. The Manx and little shearwaters (*Puffinus assimilis*) are no exception.

But the little shearwater is one of the smallest true shearwaters, with a wingspan of just 2'. Although both have dark, blackish upperparts that contrast with pale undersides, the little shearwater has more white on its face, giving it a different facial "expression." It's widespread, breeding in the North Atlantic and the oceans between South Africa and Australasia.

Manx shearwater

Little shearwater

RELATED SPECIES

• Nearly a third of all seabirds belong to the order *Procellariiformes*, including the 20 species of shearwaters, and prions, albatrosses, fulmars and petrels. The tiny Wilson's storm petrel, *Oceanites oceanicus* (below), is the most common of all seabirds.

Marabou Stork

• ORDER •
Ciconiiformes

• FAMILY •
Coniidae

• GENUS & SPECIES •
Leptoptilos crumeniferus

KEY FEATURES

● Signals dominance by inflating a gular pouch, which hangs down from the neck

● Soars effortlessly using thermals to rise high in the air

● Constructs a huge nest that balances high in the treetops

● Devours almost any animal matter, dead or alive, including carrion scraps, fish and even crocodiles and birds

WHERE IN THE WORLD?

Found in many parts of tropical Africa; from Senegal east to Ethiopia and Somalia; south through Botswana and northeastern Namibia, down to South Africa

Marabou Stork 613

LIFECYCLE

Adapted for wading, the large marabou stork also roosts and nests comfortably on branches 120' high and uses thermal soaring to fly effortlessly thousands of feet above the ground.

HABITAT

Like all birds, the marabou stork's choice of habitat is dependent upon the availability of food. Most marabou storks live in open country in arid or semiarid areas, but there is always a body of water within its flying range. The marabou hunts for fish and aquatic insects in large lakes, rivers and even small ponds. It also feeds on carrion, and for this reason, can be found scavenging near vulture attacks. With their wide variety of refuse, slaughter houses and garbage dumps also attract the hungry marabou stork. Euphorbia, acacia and baobob trees are the favorite communal roosting and nesting sites of the stork.

▼ **MADE TO WADE**
The marabou stork's long legs and partly webbed feet are well suited for wading in shallow water across its tropical African range.

▶ **FULL DEPENDENCE**
At 40 days old, the chick remains dependent on its parents for food for 90 more days.

BREEDING

The finished nest of a pair of marabou storks is impressive, measuring 3' in diameter and 1' thick. First, a 6–7 year old male selects a site, 6–120' high in a tree and waits for a courting female. She must persistently perform submissive displays because the male rebuffs both males and females approaching his territory. Once a pair is established, the male collects coarse sticks for their platform nest, which is built almost entirely by the female. She lays 2–3 eggs and both parents share incubation duties. After about 30 days, the eggs hatch at 1–3 day intervals. The hatchlings are pink and wrinkly. Both parents feed fish to the young by regurgitating food onto the floor of the nest.

? DID YOU KNOW?

- The marabou's fluffy undertail feathers, known as "marabou down," were used extensively in years past as hat ornaments.

- The marabou stork is the only species in its genus with a dark iris. The sepia-brown to grayish-brown iris gives the eye an overall brown color.

▶ **ALL TOGETHER NOW**
The silhouette of a communal roost reveals storks and their large nests.

FOOD & FEEDING

From tiny termites to colossal carrion pieces, the marabou stork eats essentially any animal matter, including baby birds. But fish and insects provide the bulk of the marabou's diet. While wading in shallow water, the marabou stork immerses its slightly opened bill, snapping it shut when it comes in contact with prey. The bird also spreads its wings slightly, luring fish into the shade, then stabbing them with its sharp bill. The marabou also consumes frogs, rodents and even young crocodiles, but it won't pass up the plant matter found among all the animal scraps in the garbage dumps it frequents.

FLAMINGO FEAST

1 In the air...
Using warm thermals for lift, the marabou stork soars high above Lake Magadi, searching for food.

2 Wall to wall...
Spotting a breeding colony of lesser flamingos with many unattended nests, the marabou descends rapidly.

3 Panic attack...
A frightened chick attempts to stand in its unguarded nest, while nervous adults watch, waiting for the inevitable.

4 Fair and square
With no parent to attempt a defense, the chick is snatched by the marabou stork, and the adult flamingos flee the scene.

BEHAVIOR

A large bird, the marabou also gathers in large numbers at communal roosts containing up to a thousand other storks. The marabou is normally silent, since its voice box has no muscles, but it does emit grunts and squeals while roosting. Perching high in the tree, it has a perfect takeoff site for its early morning flights to feeding grounds. The marabou stork uses one of the most energy-efficient ways of flying: thermal soaring. Thermals are accumulations of the early morning heat that rise, often becoming visible as tufts of cloud, which the bird easily identifies. The stork uses the rising air to ascend, then glides to the base of the next thermal. It often soars thousands of feet high, and it can be almost invisible from the ground. But it is easily distinguished from other birds in flight; its long legs are held slightly away from the body and trail behind the bird.

▶ **INFLATE TO DOMINATE**
The stork inflates its gular pouch as a sign of dominance and territoriality.

CONSERVATION

Marabou stork populations in Africa are stable. They are often tame and may be kept as pets. The marabou stork is increasing in numbers in some parts of its African range, since it appears to benefit greatly from its exploitation of urban areas with their plentiful supply of refuse. The birds now rely on garbage dumps and other sites of accumulated waste for scavenging.

PROFILE   MARABOU STORK

The marabou stork is a large bird but has almost no voice; its most formidable features are its inflated gular pouch and large pointed bill.

EARS
The stork's ear openings collect and amplify sounds from the air. The external auditory canals connect to the eardrum, which in turn passes sound to the middle and then the internal ear.

FEET & LEGS
Long legs and partly webbed feet are used for wading in the water. Because they're black, the legs and feet absorb heat and can overheat the bird. To cool off, the stork defecates on its legs and feet, turning them white.

NECK
The pinkish neck is almost bare. The gular pouch hangs 9–14" from the front of the neck. The stork inflates it, even when flying, as a sign of dominance.

VITAL STATISTICS

WEIGHT	Up to 20 lbs.
LENGTH	About 4'
WINGSPAN	7–9'
SEXUAL MATURITY	4–7 years
BREEDING SEASON	Varies according to region
NUMBER OF EGGS	1–4, usually 2–3
INCUBATION PERIOD	29–31 days
FLEDGING PERIOD	91–115 days
BREEDING INTERVAL	1 year
TYPICAL DIET	Carrion, fish, insects, birds, crocodile eggs
LIFESPAN	More than 41 years in captivity

RELATED SPECIES

● The order *Ciconiiformes* includes ibises, spoonbills and herons, as well as storks. All are wading birds, and they have three characteristics in common: long legs, long necks and large, usually pointed, bills. These features are ideally designed for capturing fish and insects while wading in shallow water. The jabiru, *Jabiru mycteria*, is the largest stork in the Americas.

CREATURE COMPARISONS

The marabou lives in Africa, while the greater adjutant stork *(Leptoptilos dubius)* lives in Asia. Habitat aside, the two birds are very similar. The greater adjutant is slate gray above, white below, much like the marabou's dark slate-gray upperparts and white underparts. Both storks stand approximately 4' tall, though the greater adjutant can grow up to 5'. A distinct feature of the marabou is its ladderlike band of white on the upper surface of the wing. This is only a single pale band of gray in the greater adjutant.

Greater adjutant stork   Marabou stork

MARTIAL EAGLE

- **ORDER** · *Falconiformes*
- **FAMILY** · *Accipitridae*
- **GENUS & SPECIES** · *Polemaetus bellicosus*

KEY FEATURES

- Largest of all the African eagles
- Extremely powerful, with the strength to kill a small antelope
- Eyesight so sharp it can spot large prey from about 3 miles away
- Hunts over an area in excess of 80 sq. miles

WHERE IN THE WORLD?

Found throughout much of Africa south of the Sahara Desert; avoids heavily forested areas and is most common (but not numerous) on savannah grasslands

LIFECYCLE

Soaring high above the African grasslands on rising currents of warm air, the mighty martial eagle is always alert, poised to dive with deadly precision on its prey.

HABITAT

The eagle's aerial view of its habitat is a rolling savannah with occasional trees dotting the landscape. It often flies over open woodland and along river valleys, hoping to spot prey unprotected by cover.

The martial eagle uses one of several favored trees for overnight roosting and regular nest sites. Important, too, is an unfailing supply of water for the regular bathing it needs.

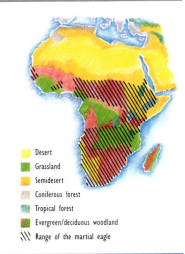

- Desert
- Grassland
- Semidesert
- Coniferous forest
- Tropical forest
- Evergreen/deciduous woodland
- \\\ Range of the martial eagle

▲ **EATING ALOFT**
Perched in the crown of a tree, the eagle uses its hooked bill to shred strips of flesh from a carcass.

▶ **KILLING FIELDS**
Open grassland is an ideal habitat for the eagle, as it offers little shelter for its prey.

CONSERVATION

The martial eagle is a rare sight because each pair of eagles ranges over a huge area in search of prey and dislikes competition from other pairs. Usually seen over savannah grassland, the eagle sometimes hunts over farmland; some farmers believe it attacks livestock and will shoot it on sight. A greater danger to this bird is habitat destruction. The use of pesticides on crops is also a grave threat, which affects the species' breeding success.

BREEDING

A female rarely lays more than one egg each season; breeding pairs invest a lot of time in their chick. For the first two months, the smaller male does most of the hunting; the female guards the nest. On his return, the male passes the food to his mate rather than directly to the chick. This lets her claim a share of the spoils.

After six months or so, the eaglet is ready to take to the air. Once it has made its first flight, it spends more and more time aloft. It flies alongside its parents, sharing their prey and learning how to hunt for itself. After a few successful kills, the young eagle leaves to find a territory of its own.

BEHAVIOR

The martial eagle hunts by sight so, like all birds of prey, it waits until the sun is up before leaving its tree roost to patrol the skies. The eagle also waits to take advantage of the sun's warmth, because the morning sun heats the ground, causing columns of warm air to rise. This broad-winged eagle is perfectly adapted for soaring on these thermals like a glider.

Launching itself into the air, the eagle flies over its domain until it senses a warm upcurrent, then starts to circle. It climbs ever higher, adjusting its slotted flight feathers to harness the wind. All the time it scans the ground for prey. If none can be found, it slips out of the thermal and glides or flies across country until it finds another thermal. The eagle may travel vast distances every day, covering a home range of 80 sq. miles or more with little effort.

▶ **TAKE-OFF**
From an acacia tree perch, the eagle launches into the sky. Once airborne, it catches a thermal and soars to a great height.

▲ **PRECIOUS YOUNG**
A female patiently shields her chick from the sun.

▶ **YOUTHFUL WARRIOR**
This juvenile isn't ready to leave the nest, but can probably kill for itself.

 FOOD & HUNTING

An eagle will attack virtually any mammal, reptile or bird it can overpower. When it selects its prey, the eagle folds its wings slightly and dives, accelerating as it rushes downward. At the last moment, it brakes hard with wings spread and thrusts its talons forward, but still strikes with enough force to tumble and even kill its prey. The eagle usually eats at the site of the kill. It may feed from a big carcass for up to five days, roosting on a nearby tree between meals.

▼ **OUT OF THE BLUE**
By sheer force of impact, a diving eagle can strike down larger prey, such as young gazelle. Smaller prey is usually killed outright by the blow.

? *DID YOU KNOW?*

● The martial eagle needs to bathe daily to keep its flight feathers clean and working efficiently.

● If an eagle attacks a young antelope, it may have to deal with its mother. The eagle spreads its wings, hisses and lashes out with its sharp talons to drive her away.

● The martial eagle's eyesight is superb. One was seen to swoop down from a hilltop and seize a guinea fowl a mile away.

SWOOPING FOR THE KILL

❶ **Reconnaissance…**
The eagle soars high above the ground, riding the breezes and relying on its superb vision to detect movement below.

❷ **Attack…**
A long, shallow dive brings the eagle swooping down on its prey. It tilts and turns in the rushing wind to adjust its course.

❸ **Strike…**
The eagle brakes sharply, and too late its prey realizes its plight: the needle-sharp talons lock forward and slam home with fatal force.

❹ **Success**
Straddling its prey, the eagle tears flesh with its hooked bill. If the carcass is small, the eagle may take it into a nearby tree.

Profile: Martial Eagle

With its rapier-sharp talons, the martial eagle makes a fearsome aerial predator, swooping down to strike its prey with terrible force.

Eyes
Large and set forward in the head with jutting, protective brows, the eyes give excellent binocular vision. They can spot even small prey at a distance of over a mile.

Wings
Broad wings equip the eagle perfectly for its hunting technique. It rises high above the plains on warm air currents, where it can soar for hours on end, conserving its energy for the moment of attack.

Legs
The legs can deliver a killer blow to prey at the moment of impact. They're fully feathered to the feet, partly for protection when the eagle strikes venomous snakes.

Tail
The tail is broad and rounded for soaring. It's also used to steer the bird during its dive and to act as an air brake just before the strike.

Feet
Powerful feet have fairly short toes tipped with sharp talons. These are lethal weapons, able to crush the life out of the prey or penetrate its body to pierce a vital organ.

Vital Statistics

Weight	11–13 lbs.
Length	32–34"
Wingspan	7–9'; female larger than male
Sexual Maturity	5–6 years
Mating Season	Varies with location
Number of Eggs	Rarely more than 1
Incubation Period	55 days
Birth Interval	1–3 years
Typical Diet	Young antelope, ground squirrels, hares, guinea fowl, snakes, lizards
Lifespan	About 16 years

Creature Comparisons

All eagles have sharp, hooked bills to tear up prey. The bill's shape reflects the eagle's diet. The bald eagle of North America has a sharply hooked bill for ripping through the scales of slippery fish. The golden eagle's bill is a little less hooked, like the martial eagle's. It's better suited to the tough skin of rabbits, hares and birds that form a large part of its diet.

Bald eagle

Golden eagle

Martial eagle

Related Species

● The martial eagle is in the family *Accipitridae*, which includes hawks, harriers, Old World vultures and kites. This is also the same family as the bald eagle and Bonelli's eagle (below) of Europe, Africa and Asia.

620 Martial Eagle

Mute Swan

- **ORDER**
  Anseriformes
- **FAMILY**
  Anatidae
- **GENUS & SPECIES**
  Cygnus olor

KEY FEATURES

- One of the heaviest flying birds, celebrated for its pure white plumage and effortless elegance
- A notoriously aggressive bird that has the power to inflict serious injuries
- Quieter than other swans, but still hisses and snorts loudly when agitated

WHERE IN THE WORLD?

Native of Eurasia; found in populations from Britain to China; introduced to North America, South Africa, Japan, Australia and New Zealand

LIFECYCLE

Portrayed as a gentle, peaceful creature in fairy tales, the majestic mute swan is actually a temperamental bird that is capable of extreme acts of aggression.

HABITAT

▲ **SWAN LAKE**
The mute swan prefers still or slow-flowing water.

Although the elegant mute swan is associated with park lakes and middle reaches of lowland rivers, most populations tend to live in wilder habitats, such as salty estuaries, coastal lagoons and marshes. Wherever it lives, the mute swan requires a plentiful supply of food, in the form of vegetation, and secure nesting sites near or in the water. However, the swan can be found nesting on industrial wastelands along visibly polluted water.

Not all mute swans are migrants, but those in northern regions, such as Scandinavia, tend to fly to temperate zones in the winter.

BREEDING

Most mute swans pair for life. They mate in spring, when a pair builds a nest together on the water's edge or in a reedbed. Once eggs are laid, the female incubates them while the male stands guard. However, the male may take over at night while his partner feeds.

Eggs all hatch at once; the downy chicks leave the nest within a day or so to swim with their parents. Although the young can feed themselves, adults use their feet to stir up food particles. At 13 weeks, chicks lose their down and obtain grayish-brown plumage; when the swans are 4–5 months old, they fly. By the following spring, most juveniles are independent and live in flocks with other non-breeding swans. Over the next two years the swans begin their search for a mate.

▼ **BRINGING UP BABIES**
Both parents care for the young, but usually drive them away after a year.

▼ **MUSIC IN THE AIR**
On a still day the rhythmic throb of the swan's slow, powerful wing beats can be heard up to half mile away.

CONSERVATION

A long association with people has made the mute swan tolerant of humans, enabling it to flourish in areas where much of its wild habitat has disappeared. It suffers from disturbance, though; some populations have been hard hit by pollution. Nevertheless, the mute swan is still thriving.

FOOD & FEEDING

Although the mute swan's main diet consists of aquatic vegetation, it occasionally extends to insects, fish, frogs and other aquatic life. In estuaries, the mute swan grazes the tender shoots of salt marsh grasses and can be seen cropping the grass in pastures near rivers and lakes. Such food is not particularly nutritious, so the mute swan spends several hours a day feeding.

When feeding in water, the swan seldom dives for its food; it plunges its head and neck below the surface to reach bottom-growing plants. With long neck and legs, the mute swan tends to feed in deeper waters than the ducks and other waterfowl in its habitat; therefore, it does not compete with them for food.

▶ **FROZEN FOOD**
Slim pickings in winters may force swans south.

BEHAVIOR

BODYGUARD

▲ **CLEARED FOR TAKEOFF**
Swans live on large bodies of water where they make long runs to get airborne.

The mute swan is bad-tempered, especially during breeding season. Each pair defends a small territory around its nest, driving away intruders with hisses and snorts. The male is particularly fierce, charging through the water with his wings raised in an awesome display of aggression. If this fails to deter an intruder, there may be a fight, with both swans beating each other with their wings. Outside the breeding season, in winter, mute swans may gather in large feeding and roosting flocks on open water with adequate food. Even here, violent conflicts between birds may occur.

Among the heaviest of flying birds, the mute swan's ascent into the air is a lengthy and spectacular show. Once airborne, however, it flies with powerful deliberation, reaching speeds of 48–54 mph. The swan's beating wings also produce a unique throbbing sound as the air is forced through the feathers.

❶ Brooding...
Incubating her eggs on a nest of sticks and reeds, the female would be vulnerable to intruders if her mate were not ready to defend her.

❷ Brave defense...
A Canada goose wandering too close finds itself in trouble as the male swan bursts out of the reeds in all his feathered glory.

❸ Beating the retreat...
The goose beats a rapid retreat with the swan in close pursuit. The swan's hooked beak and heavy wings are powerful weapons.

❹ Basking in the glory
The male returns and performs a triumph ceremony — his wings arched over his back and plumage proudly fluffed up.

❓ DID YOU KNOW?

● Mute swan's feathers were once used as quill pens. The shafts would be stripped of their filaments, baked hard in a low oven, then the nib would be cut.

● In Britain, all mute swans legally belong to the Queen.

● An adult mute swan eats up to 9 lbs. of vegetation a day.

Mute Swan 623

PROFILE MUTE SWAN

The pristine plumage and balletic beauty of the mute swan mask its true nature as a bird of enormous weight and power.

BILL

The sturdy, orange bill is slightly hooked at the tip; has serrated edges that act like a strainer when the swan is sifting food particles from the water.

JUVENILE

The juvenile mute swan gradually obtains the white plumage of the adult during its first couple of years.

FEET
Set well back on swan's body, the broad, webbed feet make for efficient swimming, but clumsy, laborious progress on land.

KNOB
Black knob at base of bill is larger in the male. It grows bigger during the breeding season to attract females.

NECK
The long neck allows the swan to reach bottom-growing plants in quite deep water by up-ending with its tail in the air.

CREATURE COMPARISONS

Named for its trumpeting call, the whooper swan (*Cygnus cygnus*) breeds on northern wetlands near the Arctic Circle; migrates south into Europe for the winter. Close in size to the mute swan, the whooper swan holds its neck erect like a goose, in contrast to the mute swan's gentle curve, and has a more wedge-shaped, yellow bill. The whooper swan has similar feeding habits to the mute swan, dabbling for water plants or grazing on wet pastures near rivers and estuaries. Like the mute swan, the whooper swan has audible wing beats, but the sound is more of a whistle than a throb.

Whooper swan

Mute swan

VITAL STATISTICS

WEIGHT	15.5–31 lbs.
LENGTH	4–5.5'
WINGSPAN	6.5–8'
SEXUAL MATURITY	3 years
BREEDING SEASON	Spring
NUMBER OF EGGS	3–12, but usually 5–7
INCUBATION PERIOD	About 36 days
FLEDGING PERIOD	120–150 days
BREEDING INTERVAL	1 year
TYPICAL DIET	Water plants, grasses and small aquatic animals
LIFESPAN	Up to 20 years

RELATED SPECIES

● Mute swan is 1 of 6 species of swan in genus *Cygnus*; also includes the trumpeter swan, *Cygnus buccinator*, of North America, and the black swan, *C. atratus* (below), of Australia. It belongs to family *Anatidae*, which contains 147 species of swan, goose and duck.

Namaqua Sandgrouse

- **ORDER** · Pterocliformes
- **FAMILY** · Pteroclidae
- **GENUS & SPECIES** · *Pterocles namaqua*

KEY FEATURES

- When faced with danger, this bird escapes by jumping from the ground directly into flight
- Male soaks water into his belly feathers, which he then takes to the nest for his young to drink
- Male incubates the eggs at night, while the female remains at the nest during the day

WHERE IN THE WORLD?

Found in southern Africa from Angola and Nambia to Zimbabwe and Botswana; also in the Kalahari Desert and South Cape Province

LIFECYCLE

The sandgrouse's activities revolve around getting water; like clockwork, flocks visit water holes to drink, and males soak their belly feathers for transport to their chicks.

HABITAT

The Namaqua sandgrouse inhabits the deserts and desert fringes of southern Africa. It prefers flat or rolling country with short, thin grass and dry, sandy soil that is covered with scattered shrubs and succulent plants. The bird also lives in more heavily wooded dry savannahs. In the Kalahari desert, the bird occurs on sandy savannahs with denser vegetation but is most common in areas that receive little rainfall. The sandgrouse will fly miles over desert terrain in search of water. If a watering hole dries up, it will journey with large flocks until another hole is found.

▲ **WIDE BUT WET**
Sandgrouse favor open areas with water holes nearby.

FOOD & FEEDING

Seeds are the main staple of the Namaqua sandgrouse, which forages over large areas — not the typical small range of most birds in the sandgrouse family. The bird prefers seeds high in protein, like legumes, picking them off the ground with its short bill. It feeds in the daylight hours and rests in the shade of a bush during the extreme heat of the midday. The bird will sometimes eat plant material and insects or mollusks that happen to be on plants, and occasionally grit to help grind down seeds in the gizzard. The Namaqua sandgrouse needs to drink water every day and may fly up to 50 miles to find a watering hole. The bird is joined by other large flocks, which all arrive at the same time: usually between 8 to 10 a.m. but sometimes in late afternoon. The birds may land as far as 5 miles from a water hole, rest from their long flight, and then walk the rest of the way. The sandgrouse will quickly take about 10 gulps, raising its head to swallow between each one.

BREEDING

Namaqua sandgrouse breed all year, but most mating occurs from July to November, when rainfall is more plentiful. During courtship, the male sandgrouse struts after the female with his tail raised and fanned and his head drawn to his shoulder. The female sometimes reciprocates with her own similar ground display. Monogamous pairs form strong pair bonds; mates stay together in small parties and sometimes in large breeding flocks. The nest is a simple scrape in bare soil or among stones, grass tufts or scrub, and the birds' nests are usually more than 60' apart.

A clutch usually consists of three eggs, though it may have two to nine; the eggs are never left unattended during the 21-day incubation period. The male incubates mainly at night from 14–18 hours, while the female's daytime duty rarely exceeds 10 hours. When the bird is relieved during incubation, it throws small stones to one side with its bill; the reason is unknown. Once the yellowish-brown chicks hatch, one parent always carries the eggshells away from the nest to avoid attracting predators, such as foxes and jackals. Chicks feed on seeds within hours of hatching, with their mother close by. Young sandgrouse are half their adult size and fully feathered at 3 weeks, and are flying at 6 weeks. The male provides chicks with water from birth until at least two months after they start to fly, since chicks usually still cannot get to the watering hole. He soaks up water in his belly feathers and returns to give it to the chicks, standing erect as they drink from a central groove in his plumage.

▲ **WHERE SANDGROUSE GATHER**
Flocks visit a water hole every morning. Some travel great distances and spend only 15 seconds getting a drink.

626 Namaqua Sandgrouse

ALWAYS INVOLVED

① Excavator...
A Namaqua sandgrouse male helps his mate form a shallow nest hollow in the very dry ground of the African desert.

② Incubator...
The sandgrouse male relieves his mate each evening at dusk and takes his turn incubating their eggs for up to 18 hours.

③ Defender...
After the chicks have hatched, the male carries the egg shells away from the nest to prevent predators from locating the chicks.

④ Dispenser
The male brings water back to the nest in his wet belly feathers. The young appear to nurse as they eagerly clamor for a refreshing drink.

CONSERVATION

Though not globally threatened, certain populations of Namaqua sandgrouse, including the Orange Free State birds, have disappeared because of severe drought and habitat destruction. But the future appears bright for the bird, due to decreased hunting and the availability of watering holes created by agricultural projects.

 BEHAVIOR

Namaqua sandgrouse are not very territorial birds, nor are they demonstrative in courtship displays. But they are very protective of their young, taking great pains to keep predators at bay. If threatened, the bird will lower its head and run at its enemy. Sandgrouse vocalizations include a mellow whistle and a chuckle, usually heard during flight. Because they are a favorite target of birds of prey, which frequent watering holes, Namaqua sandgrouse crouch in the ground to avoid detection, and can lift off quickly. In high temperatures, the bird tends to become inactive, seeking shade and drooping its wings, holding them away slightly from the body to increase heat loss.

▲ **A SANDGROUSE SPONGE**
A male transports water to his young in his feathers.

 **DID YOU KNOW?**

● The crop of a chick a few days old contained 1,400 tiny seeds — some as small as grains of sand.

● The male Namaqua sandgrouse can carry 1.5–2 oz. of water in its belly feathers. After a journey of 20 miles in half an hour, they can give their chicks 0.75–1 oz. Water loss is due mainly to evaporation.

Namaqua Sandgrouse 627

PROFILE NAMAQUA SANDGROUSE

Dry desert conditions are no threat to the Namaqua sandgrouse: its thick-soled feet can withstand hot sand, and its belly feathers carry water.

VITAL STATISTICS

Weight	5–7 oz.
Length	11"
Sexual Maturity	1 year
Breeding Season	Year-round; mainly after rains and in cooler months
Number of Eggs	2–9, usually 3
Incubation Period	21 days
Fledging Period	28 days
Breeding Interval	1 year
Typical Diet	Mainly seeds
Lifespan	Unknown

BILL
The base of the bill is feathered, which insulates the bird against extreme temperature and protects nostrils against windblown sand and dust.

MALE
A white and chestnut pectoral band and brown belly are distinguishing characteristics of the male.

FEMALE
The slightly-smaller female is mottled above with barred underparts; her belly is barred brown and white. Juveniles look similar to the female.

WINGS
With long, pointed wings, the Namaqua sandgrouse can make quick getaways from danger.

LEGS & FEET
The legs are short and three front toes are stout, fairly broad and thick-soled, well suited for walking great distances on loose sand.

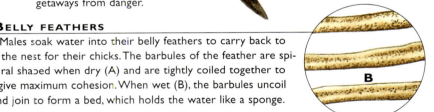

BELLY FEATHERS
Males soak water into their belly feathers to carry back to the nest for their chicks. The barbules of the feather are spiral shaped when dry (A) and are tightly coiled together to give maximum cohesion. When wet (B), the barbules uncoil and join to form a bed, which holds the water like a sponge.

CREATURE COMPARISONS

Measuring up to 16" in length and weighing up to 11 oz., the Pallas's sandgrouse (*Syrrhaptes paradoxus*) is larger than the Namaqua sandgrouse. Both females (*shown*) have mottled plumage. Unlike its relative, the Pallas's sandgrouse has no hindtoe, and its legs and feet are completely feathered, an adaptation for the cooler temperatures in its habitat. The bird lives in steppe and semidesert areas, from Kazakhstan through China, Asia, Siberia and Mongolia, far from the Namaqua sandgrouse. But the birds have very similar behavior: they both eat seeds and visit watering holes daily, and males of both species provide water to the chicks through their water-laden belly feathers.

Namaqua sandgrouse

Pallas's sandgrouse

RELATED SPECIES

- The Namaqua sandgrouse is 1 of 14 species of sandgrouse in the genus *Pterocles*, including the spotted sandgrouse, *P. senegallus*, and Madagascar sandgrouse, *P. personatus*. All have three front toes and a small, raised hindtoe. Two other species of sandgrouse in the genus *Syrrhaptes* (Greek for "sewn together") have only forward-facing toes that are fused together.

New Holland Honeyeater

• ORDER •	• FAMILY •	• GENUS & SPECIES •
Passeriformes	Meliphagidae	*Phylidonyris novaehollandiae*

KEY FEATURES

● The brush-tip tongue of the New Holland honeyeater lets it absorb all of a flower's nectar in less than a second

● A combative bird, it staunchly defends its nectar sources from other honeyeaters

● Spider webs help bind together a cup-shaped nest made mainly from bark, twigs and leaves

WHERE IN THE WORLD?

Found in southern Australia from 60–100 miles inland to the coasts of Victoria, southern Queensland, New South Wales, South Australia and Tasmania

New Holland Honeyeater 629

LIFECYCLE

Impatient and intolerant of other birds, the honeyeater stays alert to any intruders that might want to steal a sip of nectar from its favorite feeding grounds.

HABITAT

▲ **A WOODLAND WELCOME**
The varied terrain of southern Australia is home to the New Holland honeyeater.

The New Holland honeyeater establishes permanent residence in eucalyptus, banksia and melaleuca trees. The sweet nectar in the large banksia flowers makes this plant a prime choice of the honeyeater. In the bird's woodland home, trees tower up to 75', but low grassy and shrubby cover is abundant. The New Holland honeyeater also thrives in the coastal heath of southern Australia, where smaller evergreen and hard-leafed plants abound.

▶ **VERY AWARE**
Ready to defend its nectar sources.

FOOD & FEEDING

The honeyeater's remarkably efficient tongue gathers fluids from flowers in less than a second. Resembling an artist's brush at the tip, the tongue extends into the nectar about 10 times per second; the sweet nectar adheres to the licking tongue. Nectar and fruits provide a carbohydrate-rich diet for the honeyeater, but it also supplements its diet with sugary insect secretions such as honeydew, a fluid excreted by aphids. Aphids suck the sap from leaves and excrete the honeydew from their thorax onto plants, leaving the honeyeater with a sweet snack. In order to balance its meals, it also eats insects, which provide protein and other nutrients.

Several different species of honeyeater may occupy the same range, each feeding on its own favorite flowers and insects. The New Holland honeyeater will staunchly defend its favorite plants from other honeyeaters and other species of small birds. Partial to bushy flowers such as the wooly banksia, the New Holland honeyeater will feed from them until the end of their flowering cycle, then use the dried down to line its nest. Despite this preference for fluffy plants, the honeyeater isn't picky; it will feed from over 100 different flowering plant species.

BEHAVIOR

A belligerent bird, the New Holland honeyeater refuses to share its nectar supplies. It closely guards the many fluid-filled plants found in its territory from other New Holland honeyeaters and other smaller species. The bird flits restlessly from flower to flower and chatters loudly if disrupted. Honeyeaters occasionally gather for group displays, called "corroborees," but given their usual quarrelsome nature, the reason for this assembly is unknown.

The New Holland honeyeater builds extravagant nests. The basic structure is made of twigs and leaves, but it is bound together and attached to its perch with pilfered spider silk, the universal cement of bird architects. The interior of the nest is lined with the down of the dried wooly banksia flower or other similar plants. The soft, secure nest is a luxurious home for the honeyeater.

CONSERVATION

Honeyeaters are one of the most populous bird families in Australia and are currently not threatened. The wide variety of nectars and insects that the honeyeater feasts on have enabled it to thrive in a variety of habitats. Several different species of honeyeater may often occupy the same range.

 **DID YOU KNOW?**

● At least one species of honeyeater lives in every type of environment in Australia.

● The further a honeyeater extends its tongue, the less efficient the tongue becomes.

● Honeyeaters are important pollinators; they transfer pollen that clings to their feathers while flitting between flowers.

◄ ARTIST AT WORK
Like a paintbrush, the tongue collects nectar.

BREEDING

Since male honeyeaters usually outnumber females, they perform displays to attract a female during the breeding season. The male flaps in the air and emits a piping call before gliding in a steep descent to its perch. Once she has chosen a mate, the female builds a sturdy, cup-shaped nest, found 3-6' above the ground. The pinkish-buff or white eggs are laid within three days, and incubated for two weeks. Both parents feed the hungry chicks their first meals of insects, and the chicks are ready to leave the nest in about two weeks.

NESTING NEEDS

❶ Building…
The cup-shaped nest is almost complete. The female adds a few more strands of grass to finish the lining of her creation.

❷ Checking…
It has taken three days to lay a clutch of three pinkish white eggs. She inspects them before beginning her lone incubation.

❸ Gathering…
In 2 weeks the chicks have hatched. The male begins his parenting duties and collects insects for the chicks.

❹ Feeding
The male feeds the chicks their high-protein meal while the female gets ready to take her turn gathering food.

New Holland Honeyeater

PROFILE NEW HOLLAND HONEYEATER

The New Holland honeyeater is a speedy feeder as it darts from flower to flower, its tongue instantly absorbing nectar.

IN FLIGHT
Rarely stopping at any flower for more than a few seconds, the honeyeater flutters restlessly in the air.

PLUMAGE
Conspicuous black-and-white plumage appears on both the male and female honeyeater.

BILL & TONGUE
The long beak enables the honeyeater to probe a great variety of flowers, drawing out sweet nectar by the adhesive action of the brushlike tongue.

LEGS & FEET
The honeyeater's feet are designed for perching: three toes point forward and one projects behind.

VITAL STATISTICS

WEIGHT	1 oz.
LENGTH	6.5–7.5"
SEXUAL MATURITY	Unknown
BREEDING SEASON	Year-round, except February
NUMBER OF EGGS	2–3
INCUBATION PERIOD	About 2 weeks
FLEDGING PERIOD	About 2 weeks
BREEDING INTERVAL	Up to 3 times each year
TYPICAL DIET	Nectar, fruit, insects and insect secretions, including honeydew
LIFESPAN	Unknown

RELATED SPECIES

● The order *Passeriformes* contains perching birds such as broadbills, warblers, larks and honeyeaters. The honeyeater's genus, *Phylidonyris,* is one of 39 genera in the family, *Meliphagidae*. *Phylidonyris* contains six yellow-winged species, including the white-fronted honeyeater, *P. albifrons,* and the white-cheeked honeyeater, *P. nigra*.

CREATURE COMPARISONS

The white-fronted honeyeater (*Phylidonyris albifrons*) is about the same size as the New Holland honeyeater, with females smaller in both species. Both honeyeaters have touches of yellow on their wings, with the white-fronted honeyeater showing a hint of red behind its eye in contrast to the distinct white iris on the New Holland honeyeater's black face. Both birds live in woodlands, but some white-fronted honeyeaters venture into areas the New Holland honeyeater avoids, including arid shrubland and desert areas.

New Holland honeyeater

White-fronted honeyeater

NIGHTINGALE

- **ORDER**
Passeriformes

- **FAMILY**
Turdidae

- **GENUS & SPECIES**
Luscinia megarhynchos

KEY FEATURES

- One of the world's most famous songbirds; the male uses his melodious voice to attract a female
- Hardy enough to live in areas of dry shrubland, this secretive bird nests in bushy undergrowth
- Spends the winter months in Africa, returning north in summer to breed

WHERE IN THE WORLD?

Breeds in Europe, northwest Africa, the Balkans and southwest-central Asia; winters south of the Sahara, from West Africa to Uganda

Nightingale 633

LIFECYCLE

More likely to be heard than seen, the nightingale skulks in dense undergrowth singing its melodious songs during the warm nights of spring and early summer.

HABITAT

▲ **SHELTERED LIFE** Nightingales favor habitats like hedges.

The nightingale prefers low, tangled cover, which it finds in open deciduous woodland, thickets and hedgerows. Over most of its range, the nightingale is a lowland bird, but it has been recorded at 8,580' in the mountains of central Asia, and in Switzerland it can breed at 3,600'.

Many territories are near streams or pools, although the nightingale may inhabit dry hillsides in hotter parts of its range and sometimes lives in low-growing shrubs among coastal sand dunes.

DID YOU KNOW?

- When singing during the day, the nightingale frequently changes perches, but nocturnal songs are usually delivered from the same position.

- The nightingale often sings in two three-hour sessions at night, the first ending at around midnight and the second beginning early in the morning.

FOOD & FEEDING

During the breeding season, insects and other invertebrates form the nightingale's staple diet. In late summer, the bird adds berries to this intake.

▶ **OPEN SEASON** Chicks thrive on summer insects.

Leaf litter is the nightingale's favorite hunting ground. It hops about in search of ants and beetles. If there are none, it eats caterpillars, spiders and earthworms.

The nightingale may drop on prey from low branches, or pick it from the bark of its perching tree. On rare occasions, it takes winged insects, such as moths and small butterflies, from the air. Autumn brings a wealth of new feeding opportunities, and the nightingale seeks out wild cherries, elderberries, sloes and currants.

BREEDING

634 Nightingale

◀ **Pulling out**
The mother keeps the nest clean by removing the nestlings' fecal sacs.

 BEHAVIOR

The voice of a male nightingale is celebrated as one of the most beautiful of the bird kingdom. His song varies by season and circumstances. It is richest, loudest and most often performed in late spring, when the male sings at night to attract a mate and to proclaim territory. These arias are delivered from the same perches night after night, often high up and exposed. By day, his songs are less varied and are delivered in shorter bursts.

The male performs a quieter version of his song when courting a female, and maintains contact with her with short fragments of song. In the event of danger, both sexes have croaking alarm calls.

▲ **Change of tune**
The male nightingale has a wide repertoire of songs.

The nightingale selects a new partner each year, often returning to a previous site. The male returns from his overwintering grounds before the female to establish a territory. He attracts potential mates by singing and then displays at close range.

After mating, the female builds a cup-shaped nest of dead leaves and coarse grass. She incubates the eggs alone, but both parents feed the nestlings during the two weeks of rapid growth between their hatching and leaving the nest.

MAKING LOCAL CALLS

❶ Serenade...
The male nightingale stands stiffly with his wings spread as he sings to inform migrating females of his presence at a breeding territory.

❷ Court...
He lowers his voice as a female arrives, then displays to her at close range, fanning his tail and excitedly quivering his wings.

❸ Conceal...
After mating, the female collects dead leaves to make the foundation of her nest under vegetation, near or on the ground.

❹ Beg
The orange mouths of the chicks may stimulate the parent into bringing food. The most aggressive chick is often fed first.

 CONSERVATION

Like most songbirds, the nightingale suffers as habitats dwindle and pollutants enter the food chain. Numbers fluctuate year to year and vary locally: of the 10,000 pairs in the Netherlands, more are in the west than in the south and east. British populations, currently around 6,000 pairs, are declining. Overall, however, the species is not in serious decline.

Nightingale 635

Profile   Nightingale

Long legs and large feet enable the nightingale to hop among leaf litter; it is camouflaged by its reddish-brown plumage.

Eyes
The large, black eye is emphasized by a narrow ring of white feathers. The bird's eyesight is good enough to spot insects moving in leaf litter.

Juvenile
The young nightingale is mottled on the head, back and breast. It is similar to a juvenile European robin but has a reddish tail and slimmer build.

Bill & throat
The nightingale snaps up small insects in its slender bill. When the male sings, he puffs out his throat feathers.

Feet
The nightingale has typical perching feet — three toes pointing forward, one pointing backward. The foot spreads widely, enabling the bird to feed on the ground.

Vital Statistics

Weight	0.6–1 oz.
Length	6.5"
Wingspan	9–10.5"
Sexual Maturity	1 year
Breeding Season	May–June
Number of Eggs	4–5
Incubation Period	13 days
Fledging Period	11 days
Typical Diet	Insects and their larvae, spiders, fruit and berries
Lifespan	Up to 8 years

Related Species

● The nightingale is 1 of 10 species in the genus *Luscinia*. The other species tend to be more brightly colored, but their songs are not as melodic as the nightingale's. They include the Siberian rubythroat, *L. calliope*, and the bluethroat, *L. svecica*, which breeds in Europe, Scandinavia and northern Asia. *Luscinia* belongs to family *Turdidae*, which has more than 300 members worldwide.

Creature Comparisons

The nightingale's counterpart to the north and east of its range in Europe and Asia is the thrush nightingale (*Luscinia luscinia*). Unlike its relative, the thrush nightingale is restricted to lowlands and avoids dry areas. However, both birds are found in woods near water. The two species are similar-looking; even their voices are hard to tell apart, although the nightingale is the more versatile singer. The song of the nightingale is more melodic and contains fewer harsh sounds, but it is weaker than that of its relative. The thrush nightingale has obscure gray breast spots, like those of the song thrush (hence the name) and duller upperparts than the nightingale.

Nightingale

Thrush nightingale

North American Bittern

• ORDER •
Ciconiiformes

• FAMILY •
Ardeidae

• GENUS & SPECIES •
Botarus lentiginosus

KEY FEATURES

• Shy and timid, the North American bittern stalks its prey in dense marshes at night

• Brown plumage keeps it well camouflaged in marsh habitat during the day; remains motionless when threatened

• Constructs solitary nests on the ground, unlike other herons, which nest in large colonies in trees

WHERE IN THE WORLD?

Ranges in North and Central America, from south and central Canada southward throughout the U.S. and Mexico, and south to Panama; also found in Cuba and the West Indies

LIFECYCLE

The booming courtship call of the North American bittern is a common sound on warm spring evenings, as the reclusive male tries to attract a mate.

HABITAT

The North American bittern is widely distributed throughout North and Central America. It inhabits a variety of habitats, including marshes, wet meadows and bogs with fresh, brackish or saltwater. This bittern lives among tall vegetation, such as cattails, reeds, and bulrushes, relying on its streaked coloration to blend with its surroundings. Unlike other bitterns, it rarely perches in trees and is usually found on the ground. Despite being widely distributed and fairly common, the North American bittern remains out of sight; often, the only sign of its presence is its call.

▲ **WALKING TALL**
The North American bittern prefers areas with tall vegetation to escape from danger.

CONSERVATION

The North American bittern is threatened in some areas of the world. Numbers are declining in the United States, especially in the central states, due to habitat loss. The North American bittern is endangered in Illinois, Indiana and Ohio, and is designated a species of special concern in other states by the USDI (United States Department of the Interior).

BEHAVIOR

During the day, the North American bittern roosts among the reeds and rushes of the marsh. When it senses a threat, it hides by becoming completely still, pointing its bill upward and contracting its body. It blends so well with its surroundings that the intruder usually passes without spotting the bird.

▶ **WHERE'S THE BITTERN?**
The bittern's camouflage is very effective.

North American bitterns are largely migratory in the northern parts of their range. From September to November the birds migrate south, traveling only at night. They spend the winter months in the southern United States, Mexico, Central America and the West Indies. The birds return in spring.

FOOD & FEEDING

Like most bitterns, the North American bittern is largely nocturnal and ventures out at dusk to forage for food. It typically feeds alone while slowly walking in shallow water, but will occasionally feed by running after its prey. Rather than stabbing food with its sharp, pointed bill, the bittern instead grabs the food before swallowing it whole. Its diet is variable but mainly consists of aquatic prey, including eels, catfish and perch.

DID YOU KNOW?

● The bittern thrusts its head skyward in its camouflaging posture. Most birds would be left looking at the sky in this position, but the bittern's eyes are set so low, it can still see straight ahead through the reeds.

● The courtship booms of the North American bittern sound like an old pump, inspiring the common name, "thunder pumper."

BREEDING

North American bitterns are well known for their distinct booming calls during the breeding season. Males court females from April to July by calling out in the night. They fill their distended esophagus with air from the lungs and emit a distinct *pump-er-lunk* several times. The sound can carry over 0.5 mile. Males then perform displays, strutting around females with their pair of white, fanlike ruffs raised over the back.

No pair bonds are formed, and the female performs most of the parental duties. After mating, the female builds a solitary nest on a platform of dead reeds in areas of dense covering. She lays from 2–7 eggs and incubates them alone for about 30 days. The chicks fledge after about 2 weeks.

▲ GROWING FAST
The downy chicks are born helpless, but can fly in a few weeks.

SLIPPERY MEAL

① Catch...
With its hooked bill, the bittern pulls an eel from the shallow water. Gripped firmly, the eel tries to curl itself around the bird's head.

② Eat...
As the bittern swallows the eel, the feathers on the bird's head and neck become soiled from the slimy meal.

③ Collect...
After finishing its meal, the bittern collects the powdery down from its breast feathers to clean its soiled plumage.

④ Comb
The bittern then uses the fine teeth of its elongated middle toe as a comb and removes the slime from its plumage.

▼ PROTECTION
The female protects and guards her young in the nest.

North American Bittern 639

PROFILE: NORTH AMERICAN BITTERN

The North American bittern is a master of disguise, using its streaked plumage and upright pose to blend with its marshland habitat.

AIR SAC
By gulping air into its specialized esophagus and forcing it out, the bittern creates a distinctive booming call, commonly heard during breeding season.

TAIL
The bittern's tail is short and slightly rounded, an adaptation to its walking lifestyle. In flight, the long legs are held straight back and used as a rudder for steering, instead of the tail.

NESTLING
Bittern chicks are born helpless and covered with yellow-olive downy feathers.

FEET
The feet are equipped with long, unwebbed toes that give the bittern support when walking on soggy ground. The extra-long middle toe has 36 fine teeth used like a comb for grooming.

VITAL STATISTICS

WEIGHT	1–2 lbs.
LENGTH	24–34"
WINGSPAN	Up to 50"
SEXUAL MATURITY	Unknown
BREEDING SEASON	April–July
NUMBER OF EGGS	2–7; usually 3–5
INCUBATION PERIOD	28–29 days
FLEDGING PERIOD	About 2 weeks
BREEDING INTERVAL	1 year
TYPICAL DIET	Mainly fish, eels, frogs, toads, snakes and insects
LIFESPAN	Unknown

RELATED SPECIES

● The order *Ciconiiformes* contains 113 species in 38 genera and includes herons, bitterns, hamerkops, storks, shoebills, ibises and spoonbills. The North American bittern is one of 12 species of bitterns in the heron family, *Ardeidae*. There are 4 species in the genus *Botarus*, including the Eurasian bittern, *Botarus stellaris* and the Australasian bittern, *B. poiciloptilus*.

CREATURE COMPARISONS

North American bittern

South American bittern

The South American bittern (*Botarus pinnatus*) lives in Central and South America from eastern Mexico southward to Brazil. This bittern inhabits freshwater swamps and marshes among stands of tall, dense vegetation and, unlike its North American cousin, is rarely found in brackish or saltwater. Like the North American bittern, the South American bittern feeds at night on fish, eels and other vertebrates. The South American bittern appears to be a year-round resident in its range, compared to the North American bittern, which is largely migratory.

Northern Cardinal

• ORDER •
Passeriformes

• FAMILY •
Emberizidae

• GENUS & SPECIES •
Cardinalis cardinalis

KEY FEATURES

● Both male and female sing year-round and have an extensive repertoire of calls

● Bright-red plumage and bold nature make the bird a familiar sight

● The cardinal rarely migrates and usually does not wander more than a few miles from its home

● Can adapt to habitats ranging from deep forests to city gardens

WHERE IN THE WORLD?

Found in Canada, the United States from Maine to Florida and as far west as Minnesota and the western prairies in the Southwest, south to Mexico and Belize; also in Bermuda and Hawaii

Northern Cardinal 641

LIFECYCLE

The Northern cardinal displays a unique combination of bold song, color and character; it prefers to stay close to the place of its birth and rarely migrates.

HABITAT

Also known as the red-bird, the northern cardinal lives in dense thickets along field borders, in hedges, swamps, stream banks, parks and gardens. The habitat of this mainly nonmigratory bird is temperate, but its range can include desert conditions. On the arid Marias Islands off Mexico, the cardinal gets enough water by drinking the early-morning dew. The cardinal lives year-round from the Dakotas, southern Ontario, and Nova Scotia to the Gulf Coast, and from southern Texas west through Arizona and south through Mexico to Guatemala. The bird was introduced in Hawaii in 1929.

▲ **STAYING PUT**
The cardinal tolerates the cold across its range.

FOOD & FEEDING

Seeds, fruits, insects and spiders make up the cardinal's diet. In the wild, the cardinal gleans food from nearby trees and shrubs. Its wedge-shaped beak allows the bird to eat all kinds of seeds, which it holds with its grooved, upper mandible while moving the lower mandible forward to crush and husk the seed. The bird then swallows the seed's inner "meat." During fall, the cardinal ascends to tops of trees and bushes in search of grapes and berries; in the winter, the bird picks up seeds and forages around haystacks at farms. The more domesticated cardinals collect food from town gardens as well as from backyard bird feeders, favoring sunflower seeds and cracked corn. Their full menu includes 51 kinds of insects and spiders, 33 kinds of fruit and 39 types of seeds.

BEHAVIOR

The cardinal's song plays a role in all aspects of its life, from socialization to courtship to nesting. The male swells his throat, spreads his tail and drops his wing as he sways from side to side, appearing to delight at the sound of his own voice. Females begin singing their softer song in March, while males sing year-round. Both sexes defend their territories, usually a few acres in size, through the songs that they sing. The female drives out intruding females, and the males fiercely guard against other males. Songs also serve as signals, letting respective partners know when they are coming or going. The young can sing as early as 3 weeks old, but they do not aquire adult phrasing for two months.

▶ **A SPLENDID SIP**
A cardinal fans its wings as it prepares to swallow.

DID YOU KNOW?

- Northern cardinals are named for the brilliant red robes worn by the Roman Catholic cardinals.
- The cardinal is the state bird of several northern and southeastern states, including Illinois, Indiana, Kentucky, North Carolina, Ohio, Virginia and West Virginia.
- According to Cherokee legend, the northern cardinal was originally brown in color. The bird helped a wolf, which, in thanks, told the cardinal where to find a rock with red paint. The cardinal then painted himself red.

THE ART OF FEEDING

1 Follow the holes…
A male cardinal maintains a tight grip on a willow tree as it drinks from a hole previously drilled by a sapsucker.

2 Patience pays off…
A female cardinal, using her strong beak as a probing tool, discovers a large grub beneath the loose bark.

3 Easy dinner…
This cardinal pair finds its next meal without much effort on one of many bird feeders supplied by helpful humans.

4 Skilled shellers
Their broad, strong beaks manipulate the sunflower seeds; they make cracking open seeds appear almost effortless.

BREEDING

Fiercely loyal, the cardinal forms strong bonds and mates for life. During courtship, the male not only serenades his prospective mate but also feeds her. The love serenade is clear and sweet, often lasting all day. Nest building starts in March and April. The female builds a cuplike nest made of dry leaves and twigs, usually in a tree 4–5' above ground near a stream used for drinking and bathing. She adds grass and grapevines to complete the woven structure, before adding a final lining of softer materials, including rootlets and hair. It takes her about 3–9 days to complete the structure; she builds for a few hours in the morning, then a few more in the evening. The male continues his song as the female incubates the 3–4 dull white eggs with brown spots. The hatchlings are blind and helpless, with pink skin sparsely covered with gray down. Depending on the region, the cardinal pair raises from 2–4 broods during the season. The male brings insects to the young, which have large gaping mouths with red linings — easy targets when being fed. The male also guards the first nest, while the female prepares a new nest, usually about 30' away. Once the new young hatch, the juveniles of the first brood, about 3–4 weeks old, are chased from the parents' territory.

▲ **SINGLE PARENT**
The male brings food to the two hungry chicks, while the female prepares another nest.

CONSERVATION

With the northern cardinal's ability to adapt to almost any environment, this abundant species does not appear to be in any immediate danger. The bird has adapted to the Everglades of Florida and the evergreens of New York, the deserts of Mexico and suburban gardens of New Hampshire. Though sought after for the caged-bird trade in the 19th century because of its songs and brilliant plumage, the northern cardinal is currently protected. Its domestic nature, brilliant color and pleasant song make the bird welcome at bird feeders in gardens and backyards in populated areas throughout its range.

PROFILE NORTHERN CARDINAL

The northern cardinal is a familiar and welcome sight in its year-round homes, with its rich red plumage and melodic songs.

BILL
The short, wedge-shaped red bill has sharp edges, which allow the cardinal to crack open large and tough seeds. The lower mandible is broader than the upper mandible and very strong.

FEMALE
The predominantly brown female has a crest similar to the male's; her tail, however, is proportionately shorter than the male's. The base of the bill has a blackish-gray area, but it is not as noticeable as the male's.

MALE PLUMAGE
Apart from its black mask and bib, the male cardinal's colorings are eye-catching shades of scarlet. The male's brilliant red plumage is slightly glossed; the tufted crown is pointed, and can be raised and lowered at will.

FEET
The strong legs and feet are adapted mainly for perching. The feet have three toes facing forward and one behind; all toes have slender claws for support.

JUVENILE
The young cardinal resembles an adult female, but is a richer brown with a darker bill. Its crown feathers are not as long as an adult's. By the end of the first fall molt, the juvenile will attain adult plumage, but the bill-color change takes a few extra weeks.

VITAL STATISTICS

WEIGHT	1.25–2 oz.
LENGTH	7.5–8.5"
WINGSPAN	10.25–12"
SEXUAL MATURITY	1 year
BREEDING SEASON	March–August
NUMBER OF EGGS	2–5, usually 3–4
INCUBATION PERIOD	12–13 days
FLEDGING PERIOD	10–11 days
BREEDING INTERVAL	Up to 4 broods a year
TYPICAL DIET	Seeds, berries and insects
LIFESPAN	Up to 28.5 years in captivity

RELATED SPECIES

● The northern cardinal is one of several species in the genus *Cardinalis*. Its closest relative in North America is the pyrrhuloxia, *C. sinuatus*; a South American relative is the vermilion cardinal, *C. phoeniceus*. These birds are among 47 species in the family *Emberizidae*. The family includes the yellow cardinal, *Gubernatrix cristata*, as well as the red-capped cardinal, *Paroaria gularis*.

CREATURE COMPARISONS

Measuring up to 7.5", the red-crested cardinal (*Paroaria coronata*) is slightly smaller than the northern cardinal. Its gray upperparts and white underparts are accented by a large red crest on its head, throat and upper breast, quite different from the primarily red coloring of the northern cardinal. The red-crested cardinal inhabits South American savannah in Brazil, Uruguay, Paraguay, Bolivia and Argentina, where it sings from the treetops or bushes. Its path never crosses that of its northern cardinal cousin, which resides mainly in North America.

Red-crested cardinal

Northern cardinal

NORTHERN FLICKER

• ORDER •
Piciformes

• FAMILY •
Picidae

• GENUS & SPECIES •
Colaptes auratus

KEY FEATURES

• Devours more ants, whether carpenter, black or red ants, than any other North American bird

• Traps insects in a gluelike substance on the surface of its flicking tongue

• Lays an indeterminate number of eggs; if one is removed by a predator, the flicker will lay another egg to replace it

WHERE IN THE WORLD?

Widespread over North America; from Manitoba, Alaska and Newfoundland, through all mainland states south to Florida, Grand Cayman, Cuba, the West Indies and southern Mexico

LIFECYCLE

The northern flicker, which is a woodpecker, pecks on the sides of trees for food, but it mainly searches leaf litter and dead bark looking for its favorite prey — ants.

HABITAT

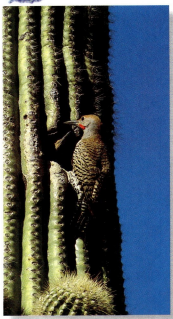

▲ THIS HOLE WILL DO
A male flicker checks out the status of his cactus nest.

The northern flicker thrives in a variety of habitats that offer open ground. From pastures to city parks to deserts, the flicker resides anywhere it can find ants, the staple of its diet. The flicker also visits farmland and orchards for fruits and berries that might be available. Roosting holes are most commonly found in trees, but barns, chimneys, and cacti will do. Nests have even been found in fence posts and haystacks.

Out in the open, the northern flicker must always be on the alert for predators, especially hawks and falcons, that prey on it. Black snakes regularly eat both the flicker's eggs and even its young.

With at least six subspecies, the flicker is widespread across the entire U.S. In the eastern states, it is the only brown-backed woodpecker. Subspecies keep to fairly distinct regions, but their ranges on occasion do overlap. Some flickers living in southern regions remain there throughout the year.

CONSERVATION

In general, populations of the northern flicker across North America are stable. However, habitat loss is a main cause of any decrease in woodpecker numbers. The European starling is the flicker's worst enemy in Colorado, competing for food and nest holes. Since woodpeckers damage trees and utility poles with their drilling, humans often target the birds.

BREEDING

At the start of the breeding season, the birds perform courtship displays including drumming, bowing and chasing in order to attract a mate. A flicker usually remains paired for life, but renews its bond each year through these rituals. After selecting a site, both birds excavate a hole that takes up to three weeks to construct. The number of eggs a flicker will lay is dependent on how many she observes in her initial clutch. If a predator robs an egg or two during egg laying, the flicker will lay replacements. The male and female share incubation duties, and an average of 5–7 chicks hatch after about 12 days. Both sexes feed the chicks until they fledge about one month later.

FOOD & FEEDING

FOUR-SEASON DINING

❶ Spring...
A mated pair digs deep in a decaying branch for wood-boring insects. One gobbles a big fat grub stuck to its tongue.

❷ Summer...
As the male searches for food for his nestlings, he discovers a nest of ants — a favorite food — hidden under some leaf litter.

646 Northern Flicker

▲ **THE WINNER TAKES ALL**
The strongest nestling climbs the nest's wall first; therefore, it is fed before its weaker sibllings.

About 45% of a northern flicker's diet consists of ants of all shapes and sizes. Termites, beetles, caterpillars, crickets and spiders make up the rest of the flicker's animal diet. It finds most insects on the ground, even searching on sidewalks for the protein-rich morsels. It's hard to escape from the gluey surface of the northern flicker's extra-long tongue. Hopping along, the bird sweeps away litter and digs into crevices and holes with its bill. The flicker also eats seeds, nuts, fruits and berries, including those of poison ivy and sumac. The flicker feeds on the ground up to 75% of the time, the only woodpecker to do so. But, like other woodpeckers, it perches on the sides of trees to glear for insects or pry out grubs, and flits between trees in search of various fruit and insect treats.

BEHAVIOR

When the northern flicker migrates to its breeding ground, the area resonates with sound. The flicker's characteristic call, *wick, wick, wick,* sounds like "wake-up, wake-up, wake-up." The call announces the flicker's arrival, and the male and female mate.

The male flicker frequently drums, especially on metal, a noise that can be annoying to many humans. But overall, the bird does not tap as much as other woodpeckers do for communication. Instead a loud *klee-yer* call is used for long-distance beckoning, especially by highly vocal fledglings.

In short flight, the flicker rapidly beats its wings to rise, then slows to dip about every 3', stalling motionlessly for a brief moment before continuing the pattern. When flickers that live in the northern regions migrate, they follow fixed courses, traveling in large flocks.

▼ **ALL ALONE**
The flicker is the only North American woodpecker that searches for food on or near the ground.

❸ **Autumn...**
Since their young have fledged, the mated pair feeds together again. Fallen cherries provide a sweet treat.

❹ **Winter**
When snow has fallen, and insects and fruit are scarce, the pair takes advantage of nuts and seeds put out by humans.

DID YOU KNOW?

- Since the flicker is such an indeterminate egg-layer, humans have removed eggs to see just how many a female flicker will lay. The record is 71 eggs in 73 days.

- Scientists have seen a male flicker treat his female partner as a rival when a fake black mustache was fastened on her face. As soon as it was removed, he accepted her back at the nest.

PROFILE NORTHERN FLICKER

The flicker is named for its flight pattern: it "flicks" up and down, revealing brilliant yellow underwings that glitter in the sunlight.

FEMALE
The female is missing the distinguishing mustache (malar stripe) found on the male. In the yellow-shafted subspecies *(Colaptes auratus luteus, below)*, the mustache is black, and both sexes of juveniles exhibit this feature. The female loses it when she matures.

BILL & TONGUE
The flicker's tongue can flick out almost 3" beyond the tip of the bill. The bird can also flick its bill, digging quickly in ant nests before capturing the insects on its sticky tongue. The bill's chisel-like shape enables it to excavate nest holes in trees.

FEET
The flicker hops on the ground and perches on branches, but unlike perching birds, it doesn't need to wrap its claws completely around a twig. With two toes forward and two toes projecting back, the foot originally designed for clinging to sides of trees allows the flicker to perch at almost any angle.

VITAL STATISTICS

WEIGHT	Up to 6 oz.
LENGTH	Up to 14"
WINGSPAN	Up to 21"
SEXUAL MATURITY	About 1 year
BREEDING SEASON	Feb.–June, varies between regions
NUMBER OF EGGS	3–12, usually 6–8
FLEDGING PERIOD	25–28 days
INCUBATION PERIOD	11–12 days
BREEDING INTERVAL	Usually 1 year
TYPICAL DIET	Mainly ants, but also other insects, fruits and berries
LIFESPAN	About 12 years

RELATED SPECIES

● The family *Picidae* has over 200 species and is found almost everywhere, with about 20 species breeding in North America. The northern flicker, *Colaptes auratus*, has many subspecies. The male yellow-shafted flicker (*C. a. luteus*) has the typical black mustache. The male red-shafted flicker (*C.a. cafer*) has a red mustache. Interbreeding can produce offspring with either black or red mustaches.

CREATURE COMPARISONS

Measuring almost 20" in length, the great slaty woodpecker (*Mulleripicus pulverulentus*) is 40% longer than the northern flicker, and at 19 oz. weighs three times as much. It is the largest Old World woodpecker. The great slaty is slate-gray (hence its name) and has a long neck and tail, perfect for long days spent hammering and drilling on the sides of trees in Southeast Asia. This sharply contrasts with the varied colors of the flicker's plumage and its shorter neck, adapted for life spent mainly on the ground in North America foraging for ants and other insects.

Northern flicker   Great slaty woodpecker

Northern Gannet

• ORDER •
Pelecaniformes

• FAMILY •
Sulidae

• GENUS & SPECIES •
Sula bassana

KEY FEATURES

• Glides over the ocean for hours on end, "riding" the rising currents of air deflected by the waves

• Hits the sea at 60 mph when diving after shoals of fish

• Stores its catch in a throat pouch until it's ready to be swallowed

• Before it can fly, the single young leaves its nest and starts to swim south for the winter

WHERE IN THE WORLD?

Confined to the northern Atlantic, breeding on the coasts of Iceland, Norway, France, Britain, Ireland and northeast Canada; winters at sea, south as far as the coasts of northern Africa and the southeastern U.S.

Northern Gannet 649

LIFECYCLE

The northern gannet spends much of its life in dashing flight at sea. But at its breeding colonies on land, it occupies its time in ritualized displays with its mate and neighbors.

CONSERVATION

In the last century, the northern gannet was killed for food, but legal protection ended this practice and it recovered. A recent threat is overfishing of the fish stocks on which it feeds.

HABITAT

◀ **WHERE THE WIND BLOWS**
Air rising against cliffs helps the heavy gannet take off on its 6' wings.

The gannet wanders over open seas, often up to 90 miles from the coast, coming to land only to breed. It nests on offshore islands with steep cliffs in colonies known as *gannetries*. From afar, a gannet island seems to be covered in snow because of the white birds and their *guano* (droppings).

Island-nesting has great advantages: the gannet is safe from predatory mammals and can fly in any direction in search of fish. It tracks seasonal movements of fish shoals, moving south to warmer parts of the Atlantic in autumn and returning in spring.

FOOD & FEEDING

▼ **DIVE BOMBER**
The gannet often plunges nearly 100' to catch fish.

The gannet travels great distances to find rich feeding grounds. It spots shoals of fish while flying high over the sea, probably by the iridescent "oil slicks" shimmering on the surface that rise from the shoals. Then, arrowlike, the gannet dives vertically with its wings partly closed and plunges deep with a huge splash.

The gannet's weight and diving speed thrusts it about 12' below the surface, but it can paddle down to 50' in pursuit of fish. It can even catch strong-swimming fish, such as mackerel. Small prey is swallowed at the surface before the gannet flies up to dive again and again.

DID YOU KNOW?

● Almost three-quarters of all northern gannets, about 200,000 pairs, breed off the coasts of Britain and Ireland.

● Male gannets often have yellow stripes on the toes; females may have blue.

● The gannet can swallow four large mackerel one after another and may become too heavy to fly up from the sea's surface.

● It takes 4–5 years for a juvenile to gain adult plumage, passing through ever-whiter stages.

BEHAVIOR

The gannet nests in huge colonies, and all the birds look almost identical; there's little difference between sexes. It therefore uses a ritualized "language" of postures and calls to find its mate and keep neighbors at a distance.

A male claims ownership of his nest by calling loudly every time he returns to the colony and by displaying to neighboring males. He jabs the air with his bill, bows with wings raised, gapes menacingly and grapples bill-to-bill with rivals. Pairs communicate in gentler displays, such as bill fencing.

650 Northern Gannet

BREEDING

◀ **MESS OF A NEST**
Seaweed, feathers and *guano* make up the nest.

▼ **EARLY BIRDS**
At dawn, the colony is already active.

At the start of the breeding season, the male brings nest material to his mate, which she binds together with droppings. They incubate their egg in alternating 30–36-hour sessions for 43 days.

The downy offspring is fed by both parents for 13 weeks. Too fat to fly and its wings not fully grown, it jumps into the sea to head south for winter by swimming and drifting with the current. It survives on fat reserves until it can fly and fish for itself.

▼ **BILL RATTLING**
When a bird returns to the nest, it engages its mate in friendly "fencing."

SIGN LANGUAGE

❶ Finding the way…
A gannet colony is crowded, with thousands of similar-looking nests all only 3' apart; an incoming bird uses landmarks to find its nest.

❷ Token of esteem…
A male arrives with seaweed for his mate to add to the nest. The birds pair for life and use the same nest each year.

❸ Good to see you…
Shaking their heads up and down and from side to side, as if in slow motion, the birds strengthen their bond.

❹ See you soon
One bird points to the sky to warn its mate that it's about to leave. If both birds left, their egg or chick would be at great risk.

PROFILE NORTHERN GANNET

The northern gannet is equipped for life as a specialized fish-eater and flies great distances on its long wings to track down prey.

JUVENILE
The juvenile's plumage is dark brown, maybe to signal to aggressive adult males that it isn't a rival. In noisy disputes, adults often attack each other, even their own mates, on occasion.

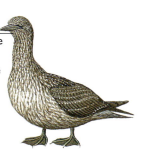

BILL
Tiny serrations along the cutting edges of the mandibles act like the teeth of a fine saw, slicing into the skin of squirming fish to hold them securely.

THROAT POUCH
Large fish are held in an expandable throat pouch until they're ready to be fully swallowed. As many as ten fish have been found in the throat pouches of some gannets.

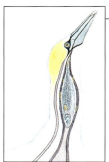

FEET
All four toes of each foot are webbed, providing power to drive through the water in pursuit of fish. During incubation, the gannet wraps its feet around its egg, like a blanket, to keep it warm.

VITAL STATISTICS

WEIGHT	5–8 lbs.
LENGTH	3–3.3'
WINGSPAN	5.5–6'
SEXUAL MATURITY	4–5 years
BREEDING SEASON	April–September
NUMBER OF EGGS	1
INCUBATION PERIOD	43 days
FLEDGING PERIOD	91 days
BREEDING INTERVAL	1 year
TYPICAL DIET	Fish: mainly shoaling species, such as herring and mackerel
LIFESPAN	15 years

RELATED SPECIES

● There are 2 other species of gannet in the genus *Sula*: the Australian gannet, *S. serrator*, and the Cape, or African, gannet, *S. capensis*. Both fish the same way as the northern gannet, but in temperate oceans in the Southern Hemisphere. *Sula* also has 6 booby species: the blue-footed, red-footed, brown, masked, Peruvian and Abbott's boobies. These birds take the place of gannets in tropical seas.

CREATURE COMPARISONS

Smaller than its relative the northern gannet, the masked booby (*Sula dactylatra*) or the blue-faced booby, ranges over all tropical oceans and nests on the flat, open ground of tropical islands. The booby is lighter than the northern gannet, which lets this long-winged bird take off from islands where there's often very little wind to provide lift.

Both species have a mainly brilliant white plumage, but the northern gannet has creamy-yellow feathers on its head. The masked booby also has black trailing edges on its wings and tail, whereas the gannet is black on its wing tips alone.

Northern gannet

Masked booby

Northern Goshawk

- **ORDER** · Falconiformes
- **FAMILY** · Accipitridae
- **GENUS & SPECIES** · *Accipiter gentilis*

KEY FEATURES

- Highly aggressive, it quickly kills its prey with the vicelike grip of its powerful talons

- Fiercely defends its nest from intruders, even attacking humans who venture too close

- Extremely acute vision allows it to quickly find and capture prey

- There are about eight subspecies of goshawk

WHERE IN THE WORLD?

Found in North America, from Alaska and Canada south through most of the United States and parts of Mexico; also found in parts of Asia, Africa and throughout most of Europe

LIFECYCLE

The northern goshawk leads a mainly solitary life, often surveying hundreds of acres and feeding on a wide variety of prey, from grouse to grasshoppers.

HABITAT

The northern goshawk is widely distributed throughout the temperate regions of the world, including North America, Europe and parts of Asia and Africa. This hawk mostly inhabits mature woods with remote stands of tall timber, particularly coniferous forests, but also deciduous and mixed forests. Found from sea level in lowlands to the subalpine woods of mountainous regions throughout its range, the northern goshawk prefers dense forests adjacent to large clearings, where it can hunt a wide variety of prey.

▼ **OPEN HOUSE**
The goshawk prefers forests close to open fields for hunting food.

CONSERVATION

The northern goshawk is not globally threatened, but is declining in many parts of its range. Habitat destruction and deforestation have caused significant declines, particularly in western Europe. Like all raptors, the goshawk is vulnerable to pollution, especially from pesticides and heavy metals that cause thinning of its eggshells and a decrease in its prey. The goshawk's nest sometimes is robbed to provide birds for the sport of falconry. It is protected by law in the U.S. and is listed on Appendix II of CITES (Convention on International Trade in Endangered Species).

FOOD & HUNTING

Highly aggressive, the northern goshawk feeds on a wide variety of prey. Its diet mainly consists of small and medium-sized birds and mammals, including hares, rabbits, squirrels, chipmunks, weasles, ducks, grouse, quail, owls, small hawks and woodpeckers; it also eats insects, such as grasshoppers and caterpillars. Hunting from a hidden perch or surveying a large clearing while in flight, the northern goshawk swoops down on unsuspecting prey and catches it on the ground; it may also skillfully dart and twist through the dense forest at speeds up to 38 mph in pursuit of birds. Once its victim is caught, the northern goshawk takes it to a clearing, devouring it on the ground. Like most hawks, the northern goshawk performs a mantling display when feeding in order to guard its hard-earned meal from other birds of prey that might try to steal it. The bird spreads its wings over its prey to create a canopy while puffing up its feathers and making threat calls.

▶ **ON THE MANTLE**
The goshawk protects its food from thieves with a mantling display.

HIDDEN PERCH

❶ **Perching...**
From a hidden perch high in a tree, a northern goshawk looks out over an adjacent clearing and surveys it for prey.

BEHAVIOR

The northern goshawk is solitary except during the breeding season. Although it is active all day, it usually hunts during the early morning and evening. During the hottest parts of the day, the northern goshawk roosts in a tall tree where it watches over its territory. Like all raptors, the goshawk is highly territorial and defends an area as large as 5,000 acres in some parts of its range. The northern goshawk is mainly sedentary, but some populations in the northern parts of its range migrate as far south as Mexico in October for the winter months; these populations return to the northern breeding grounds in March.

▲ **SOARING HIGH**
The goshawk majestically soars through the air.

DID YOU KNOW?

● The goshawk is very bold when hungry and has been known to seize chickens in the presence of humans; it has even attacked wooden duck decoys that were placed in a marsh by hunters.

● When hunting or angry, the goshawk's eyes turn dark crimson; the eyes of an angry female were such a fiery blood red that an observer thought she was seriously injured.

● The female frequently pulls and loosens the bottom of the nest; this provides sufficient aeration to prevent the growth of mold and maggots.

② Swooping...
Spotting a snowshoe hare (*Lepus americanus*), the northern goshawk swoops down from its perch toward its victim.

③ Seizing...
Seeing the goshawk, the hare attempts to escape by racing away; however, it is no match for the hawk's powerful talons.

④ Holding
With the vicelike grip of its sharp talons, the goshawk firmly holds onto the struggling hare.

 ## BREEDING

Like most birds of prey, the northern goshawk is monogamous and mates for life. After wintering alone, the female returns to the nesting site from the previous year and calls to attract her mate with high-pitched screams. Both sexes perform flight displays, which include soaring over the nest site. The nest, made of sticks lined with twigs and leaves, is repaired, or a new one is built. After mating, the female lays 1–5 eggs (usually 3–4) and does most of the incubating, turning the eggs every half hour. The male hunts and brings her food during this time. After 35–38 days, the eggs hatch and the chicks are covered in grayish-white down. Both sexes vigorously defend the nest from intruders, including humans, and attack anything that ventures too close. The young hawks fledge in 34–45 days, with the males fledging about one week before the females; however, they do not permanently leave the nest for 70–90 days. Sexual maturity occurs between 2–3 years of age, but sometimes female goshawks mate at 1 year.

▲ **PARENTAL CARE**
Well known for its attacks on intruders, the northern goshawk protects and cares for its young.

PROFILE: NORTHERN GOSHAWK

With its short, rounded wings and long, narrow tail, the goshawk twists and darts at incredible speeds through the forest in pursuit of prey.

EYES
Similar to mammals, the hawk's eyes have a lens (A), cornea (B), iris (C) and retina (D). The long distance from lens to retina gives the hawk excellent vision, allowing it to see a very detailed and colorful world.

BILL
The sharp, hooked bill rips into prey. The nostrils are covered by an area of bare skin called the *cere*.

WINGS
The short, rounded wings allow the goshawk to make sharp turns and dodge obstacles, such as tree branches, as it weaves skillfully through the trees.

PLUMAGE
All subspecies of goshawk have barred plumage that allows them to blend with their surroundings while they hunt.

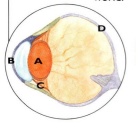

WHITE MORPH
Found in Russia from Siberia to Kamchatka, the subspecies *A. g. albidus* may be light gray with barring on its chest, or white with only faint barring.

Subspecies shown: northern goshawk, *Accipiter gentilis actricapillus*

VITAL STATISTICS

WEIGHT	Male 1–2.5 lbs.; female 1.75–3.5 lbs.
LENGTH	19–27"
WINGSPAN	38–50"
SEXUAL MATURITY	2–3 years
BREEDING SEASON	April–June
NUMBER OF EGGS	1–5; usually 3–4
INCUBATION PERIOD	35–38 days
FLEDGING PERIOD	34–45 days
BREEDING INTERVAL	1 year
TYPICAL DIET	Mostly small birds, mammals and reptiles
LIFESPAN	Up to 19 years in captivity

RELATED SPECIES

● The northern goshawk is the largest member of the 50 species in the genus *Accipiter*; close relatives include the African goshawk, *A. tachiro*, the Chinese goshawk, *A. soloensis* and the Cooper's hawk, *A. cooperii*. The family Accipitridae contains 237 species in 64 genera of hawk and eagle. There are about 289 species of hawk, eagle, falcon and vulture in the order Falconiformes.

CREATURE COMPARISONS

Found only in Cuba, the Gundlach's hawk (*Accipiter gundlachi*) is a close relative of the northern goshawk. Unlike the goshawk, which inhabits the mature coniferous and deciduous forests of lowlands and mountainous regions, the Gundlach's hawk occurs in the tropical, evergreen and marshy forests along the coastline. The Gundlach's hawk is dark gray above with a gray breast and rufous barring on its belly and thighs, compared to the goshawk, which has a blue-gray back and white and gray underparts; both species have black caps on their heads. At 16.5–20" in length, the Gundlach's hawk is smaller than the northern goshawk.

Northern goshawk   Gundlach's hawk

Northern Harrier

- ORDER -
Falconiformes

- FAMILY -
Accipitridae

- GENUS & SPECIES -
Circus cyaneus

KEY FEATURES

● An elegant bird of prey that hunts at a leisurely pace, flying low over open country in search of small birds and mammals

● In courtship aerobatics, the male tumbles in display or drops food to the female in a midair pass

● Once severely persecuted in Britain, communal winter roosts are now widespread again

WHERE IN THE WORLD?

Breeds throughout North America, Europe and Siberia; winters in southern U.S. south to South America; in Europe from southern Asia and southern Europe to northern Africa

LIFECYCLE

A methodical hunter, the northern harrier turns into a stunt flier as it launches into a spectacular sky dance to impress its mate.

HABITAT

A bird of wide open spaces, the northern harrier is widespread across the U.S. and Canada. It prefers to breed in marsh and wetland edges; it was in this habitat that the bird was given its old name of marsh hawk. In Europe and Asia, where it is called the hen harrier, it breeds in large expanses of heather, gorse (a thorny bush) and boggy hollows, but also in the moutains up to 8,000'. During the winter, the bird migrates to warmer areas where it prefers fields, pastures and coastal marshes.

▼ **HARRIER HAUNTS**
The northern harrier prefers covered areas.

FOOD & HUNTING

Unlike most birds of prey, the northern harrier relies on slowness rather than speed when hunting. Its wings and tail are large in relation to its slim body, enabling it to fly at speeds of less than 12 mph as it scans the ground from a height of 10–13'. The bird adjusts its speed according to its target and terrain. It is at its slowest when hunting voles on rough ground, but the harrier reaches speeds of 25 mph when skimming over ploughed fields or short grass in search of birds. The northern harrier also relies on its acute sense of hearing to locate prey. It has a facial disk of feathers which helps to channel the faintest of squeaks into its large ears.

Once it has located a prey item, the northern harrier suddenly pounces feet first on the intended victim. It may kill only once in every 15 attempts.

DINING OUT IN EUROPE

❶ **Scouting mission...**
A male harrier hunts low over a Scottish moor, scanning the ground and listening carefully for prey.

❸ **In for the kill...**
The northern harrier pounces on the unsuspecting grouse, stabbing the bird lethally with his talons.

❓ DID YOU KNOW?

● The northern harrier spends about 40% of daylight hours on the wing and covers about 100 miles every day.

● Harrier breeding pairs often dive-bomb people who venture too close to their nests. They may also strike with their talons.

BREEDING

▲ **INCUBATION STATION**
The harrier's nest is built out of sticks, heather, gorse and grass.

Aerial courtship displays by the northern harrier end in a sky dance, in which the male climbs to a height of 1,600' before tumbling down. He may plunge and rise more than 20 times before landing. Northern harrier pairs also execute "food passes." The male bird drops the food, and the female rolls onto her back, catching it in an outstretched foot.

Each female lays between three and six eggs, which she incubates for about a month. Once the chicks hatch, she feeds them on morsels of meat brought to her by the male. The young harriers fledge a month after they are born.

▶ **FEED ME!**
Chicks rely on parents for food for over a month.

② Weakened defense...
He spots a red grouse feeding in the heather below, and swoops swiftly to the attack.

④ Filling feast
The prey is slowly plucked and eaten at the site where it was captured.

BEHAVIOR

During the winter, northern harriers in Europe and Africa will often gather at traditional communal roosts. The groups average about 30 birds, but more than 200 birds sometimes assemble at these sites. The birds circle the roost together at dusk before settling down for the night.

Roosts are located on the ground in boggy or flooded areas. This offers protection from foxes and other enemies. The bird sleeps on a "bed" of trampled grass or heather.

Communal roosting may provide the harrier with a food-finding system. A hungry harrier can recognize a better-fed individual and will follow it to a more productive hunting area the following morning.

Away from its roost, the northern harrier is generally solitary, but it may travel in small groups on autumn migration.

The northern harrier's vocal range is limited to cackles and wails. It is silent outside the breeding season, except when disturbed at its roost.

▲ SHOCK TACTICS
Young northern harriers screech noisily and flap their wings when startled.

 CONSERVATION

Like all birds of prey in the U.S., the northern harrier is protected by law. It is legally protected in Britain as well but, in some European countries, the bird is still shot, especially during its migration. This can hurt local populations.

Northern Harrier 659

PROFILE   NORTHERN HARRIER

The northern harrier's light build and long wings give it buoyant, energy-efficient flight, enabling it to soar for long periods as it hunts.

MALE
The male is distinguished from the female by his gray plumage. In flight, he reveals black primaries and a dark band along the trailing edges of his wing undersides.

CERE
The cere, a waxy, fleshy covering at the base of the bill, is yellow in the adult but greenish in the juvenile.

FEMALE
The female harrier has brown plumage with a streaked breast and white rump.

TAIL
The long tail aids maneuverability when the bird is hunting. Banding on the female's tail accounts for her common name: "ringtail."

FEET
Short toes and needle-sharp talons at the end of long legs pin prey to the ground.

VITAL STATISTICS

WEIGHT	Male 12 oz.; female 18 oz.
LENGTH	17–23"
WINGSPAN	38–48"
SEXUAL MATURITY	2–3 years
BREEDING SEASON	April–June
NUMBER OF EGGS	3–6
INCUBATION PERIOD	29–31 days
FLEDGING PERIOD	32–42 days
BREEDING INTERVAL	1 year
TYPICAL DIET	Small birds and rodents
LIFESPAN	Up to 16 years; usually less than 12 years

CREATURE COMPARISONS

The marsh harrier (*Circus aeruginosus*), which is widely distributed in Europe, Africa and Asia, is a relative of the northern harrier. The male marsh harrier is mainly dark chestnut-brown with gray wings and tail, while the female is chocolate-brown with a yellowish-cream head, throat and forewing. Like the northern harrier, the marsh harrier hunts by flying low over the ground, but does so at a greater speed. Its wingbeats are heavier, too.

Male northern harrier   Male marsh harrier

RELATED SPECIES

● Harriers, as a group, are found virtually throughout the world, but expert opinion on the number of species ranges from 9–13. Montagu's harrier, *Circus pygargus* (below), of Europe, looks much like the northern harrier.

660 Northern Harrier

Northern Mockingbird

- **ORDER** · Passeriformes
- **FAMILY** · Mimidae
- **GENUS & SPECIES** · *Mimus polyglottos*

KEY FEATURES

- Mimics about 30 different species of bird and a variety of other sounds, from whistles to barks
- Boldly defends a wide territory but reduces its area size in winter
- Eats fruits and insects; devours a large number of harmful crop pests

WHERE IN THE WORLD?

Widely distributed throughout North America, from Alberta and Quebec in Canada south to Mexico and the Bahamas

LIFECYCLE

The northern mockingbird, a master mocker, deceives hunters and birdwatchers with its song imitations, and battles intruders in mock fights and pecking matches.

HABITAT

▲ **OUT OF THE WAY**
The cup-shaped tree nest conceals the newborns.

▼ **NAME THAT TUNE**
A juvenile tries out a new imitation at its tree perch.

Found throughout North America, northern mockingbirds in the northernmost areas migrate south for the winter. But in the rest of its range, the mockingbird sets up permanent residence in a variety of open habitats. From lawns and farms to gardens and pastures, the northern mockingbird is found wherever there is an attractive mix of trees, shrubs and grasses. This blend of vegetation provides food and protection for the mockingbird. Year-round homes include streamside thickets and schoolyards in the north, orchards and vineyards in the west and even cactuses in the southwest. The mockingbird has even been introduced and is a year-round resident in Hawaii. In the winter, the bird stakes out and boldly defends a small territory centered around a food source. But with the start of the breeding season in January, the male mockingbird establishes a much larger nesting territory in preparation for starting a new family.

BREEDING

The male signals the start of the breeding season with a loud, brash call. When a female enters its territory, the two birds watch each other and communicate with a loud series of calls. Once a pair bonds, the frequency of their songs decrease. The birds work quickly to construct a strong, cup-shaped nest of twigs, grasses and rootlets about 3–10' above ground. The female lays up to six eggs and begins incubating after laying the last one. The hatchlings are fed by both parents for about 12 days and then fly off to establish a territory of their own. More than one brood can be raised during the breeding season.

▶ **A HUNGRY TRIO**
With mouths open, young hatchlings resemble bright yellow flowers.

BEHAVIOR

The mockingbird has an uncanny ability to mimic at least 30 other birds' songs. The imitations are incredibly precise. For example, the red-winged blackbird can't tell for sure if a song is from a fellow blackbird, indicating that the territory is already claimed. This mimicking helps keep other species out of the mockingbird's territory. When establishing its domain, the mockingbird loudly repeats a note at least three times, then switches to another phrase or imitation. The mockingbird's repertoire is not limited to bird songs. It can imitate the tinkling of a piano, squeaky hinges and even a dog's bark. When its songs don't warn off trespassers, its aggressiveness is a quick deterrent. Dogs, cats and squirrels are savagely pecked. Mockingbirds can be harsh with other birds as well; it assumes a threatening posture, with tail cocked and fanned, against robins, starlings and woodpeckers that dare try to rob from its area.

COMBAT ZONES

❶ **Backyard guard...**
The mockingbird maintains its breeding territory with an ever-changing song. It scans for intruders from its treetop perch.

❷ **Battle-scarred...**
The bird considers the suburban garden as its domain. It swoops down for attack and pecks madly at the unwelcome resident dog.

DID YOU KNOW?

- Chosen for its singing abilities or territorial allegiance, the mockingbird is the state bird for Arkansas, Florida, Texas, Mississippi and Tennessee.
- In the Pueblo culture, it is believed that the mockingbird grants the gift of speech to humans.

FOOD & FEEDING

The northern mockingbird eats a variety of fruits and insects. Domestic fruits such as grapes, oranges, blackberries and figs account for a portion of the bird's diet, but wild fruits such as holly, smilax, elderberry, mulberry and fruit of the prickly pear cactus account for about 45% of the total food eaten. Almost half of its daily diet is insects, including harmful crop pests such as boll weevils, cucumber beetles and chinch bugs. Grasshoppers, caterpillars, spiders, sow bugs and snails round out the high-protein portion of the mockingbird's meal. The bird pounces on bugs from its perch with a swift descent, often catching an insect in midair above a tangle of wildflowers. The mockingbird also takes advantage of food such as bread, raisins or suet (hard beef fat) left out by humans. This easy snack is welcome, especially during the cooler winter months, when food is scarce in northern regions where birds remain year-round.

▲ **A BITTERSWEET FEAST Berries and fruits are the staples of the northern mockingbird's diet.**

❸ Rite of spring...
Boundary disputes occur between March and August when males confront each other at the edge of the breeding territory.

❹ Boxing ring
They look like boxers at the start of a match, but only hop along a disputed border until the intruder retreats in defeat.

CONSERVATION

The northern mockingbird's affinity for crops of sweet fruits and berries may put it in jeopardy. One grape grower killed 1,100 mockingbirds to keep them from eating portions of his valuable crop. But the northern mockingbird has adapted well to urbanization of its natural habitat, and populations remain strong. Habitats modified by open tree and shrub plantings simulate the natural forest-edge conditions preferred by the mockingbird and have led to its spread across the United States. The push northward has been slowed by destruction of similar woody habitats.

Northern Mockingbird 663

PROFILE NORTHERN MOCKINGBIRD

Despite its dull plumage, the northern mockingbird attracts attention with its mimicking medleys and daredevil territorial defense tactics.

JUVENILE
The brown juvenile has faint spots on its chest that effectively conceal the bird from predators.

FEET
The long feet are perfect for perching in treetops, from which the mockingbird establishes its territory. But powerful legs and feet also support the bird when it hops through the undergrowth in search of food.

BILL
The short, slender bill is ideal for picking up berries or insects. But the mockingbird also uses the pointed tip to peck at dogs, cats and even humans who dare invade its space.

VITAL STATISTICS

WEIGHT	1.5–2 oz.
LENGTH	9–11"
WINGSPAN	13–15"
BREEDING SEASON	March–August
NUMBER OF EGGS	3–6, usually 4–5
INCUBATION PERIOD	12–14 days
FLEDGING PERIOD	10–12 days
BREEDING INTERVAL	2–3 broods per season
TYPICAL DIET	Berries, seeds and insects
LIFESPAN	In captivity, up to 15.5 years; in the wild, about 12 years

RELATED SPECIES

● A member of the family *Mimidae*, the northern mockingbird joins about 13 other species of mockingbird in this group. The blue and white mockingbird, *Melanotis hypoleucos*, is a vibrant, colorful contrast to the typical brown or gray of most mockingbirds. Approximately 17 species of catbirds and tremblers are relatives included in the *Mimidae* family of singers and mimics.

CREATURE COMPARISONS

The 10" long Galapagos mockingbird (*Nesomimus trifasciatus*) is comparable in size to the northern mockingbird. But the Galapagos mockingbird, as its name suggests, is found only on the Galapagos Islands of Santa Fe, Isabela, Fernandina and Darwin. There, they form family flocks, in contrast to the ess gregarious northern mockingbird. Both species eat insects, seeds and berries. But the Galapagos also scavenges for other fare, including sea lion placentas and the eggs and even chicks of other birds.

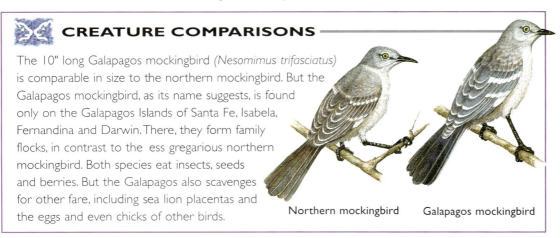

Northern mockingbird   Galapagos mockingbird

664 Northern Mockingbird

Northern Pintail

• ORDER •
Anseriformes

• FAMILY •
Anatidae

• GENUS & SPECIES •
Anas acuta

KEY FEATURES

- Most widely distributed duck in North America; one of the most numerous ducks in the world
- Averages over 100 miles per day on long migratory flights that often exceed 2,000 miles
- Named for its "pintail," the elongated tail feathers that can measure up to 4" long in males

WHERE IN THE WORLD?

Found throughout the Northern Hemisphere; breeds across Canada, the U.S., Europe and Asia; also winters in South America, Africa, and India

LIFECYCLE

Female and male pintails resemble each other when the male molts; when his new handsome plumage grows in, he migrates south a month ahead of the female.

HABITAT

From the tundra to the tropics, the northern pintail is widespread and numerous The pintail's two basic requirements for any home are water, shallow enough for feeding, and low vegetation on the shores. It is most abundant on prairie and tundra habitats that offer open vistas with quiet and shallow marshes, rivers or lakes. Freshwater is preferred over brackish. In Alaska, thousands of small ponds are home to one or more pairs. Pintails usually remain on ponds near the coasts, but in India, large numbers are found on inland lakes. On the sea, the birds form flocks of over 1,000, which split up into small parties that mingle with other ducks.

▼ **MOVING DAY**
Millions of pintails migrate each spring and fall.

BEHAVIOR

▲ **AQUATIC RUNWAY**
The pintail has large wings for strong, fast flight.

From mid-July to early September, the male pintail, or drake, molts, trading his handsome plumage for a plainer set of feathers called eclipse plumage. The pintail molts all of its flight feathers simultaneously, rendering it flightless. To compensate, the male assumes the overall coloration of the camouflaged female. During this vulnerable flightless period, a pintail will only dive fully underwater in an attempt to escape danger. The males molt first into their winter plumage and begin their long migratory journey south; the females follow a month later. The pintail accurately retraces the same migratory flyways in the fall that it followed in the spring. Flight speeds reach over 50 mph, but the young have less endurance and require more frequent stops than the adults. Only those pintails in parts of the western U.S. and Europe remain in the same place throughout the winter.

▼ **PERFECT PLUMAGE**
The pintail drake's tastefully colored feathers form an elegant profile.

CONSERVATION

The pintail is not endangered; over 6 million pintail breed in the U.S. alone. But hunting and drainage are contributing to a slow decline of many ducks. The pintail is a popular target for hunters; in fact, it is among the top three, along with mallards and teal. The draining of marshes and shallow lakes is continuing at an alarming rate, reducing breeding and nesting grounds.

FOOD & FEEDING

The northern pintail is an opportunistic feeder, using a variety of techniques to exploit all that its shallow, aquatic habitat has to offer. The pintail's flexible neck, serrated bill and webbed feet enable it to skillfully swim across shallow waters and surface-feed on seeds, grasses, insects and tadpoles. But the long neck offers an additional payoff: the pintail can feed in relatively deep waters by up-ending. Powerful, paddling legs help the duck keep its balance, while it reaches up to 12" below the surface. Spending up to 6 seconds underwater, the pintail searches for snails, mollusks, crabs and seeds along the muddy bottom and often uproots pondweed and sedge plants.

FLEXIBLE FEEDERS

❶ Dabbling...
The pintail is a dabbling (surface-feeding) duck. A female sieves aquatic plant matter with the help of serrations on her bill.

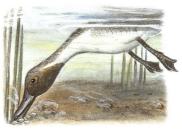

❷ Head-dipping...
The pintail can also extend its neck into the shallow waters to reach various submerged plants, found just below the surface.

❸ Up-ending...
To reach the tubers of aquatic plants, the female submerges the front of her body, while her tail remains visible above the surface.

❹ Grazing
The pintail's feet are set farther forward than those of diving ducks, so it can wander comfortably over nearby pasture.

 ## BREEDING

The most common male courtship display is a call known as "burping"—a geee sound that changes to a soft, flutelike whistle. Males burp,

▶ **DOWNY SOFT**
All chicks hatch together after an incubation of about 23 days.

then perform a "head-up-tail-up" display, with their long tails pointing straight up. Once she has chosen a mate, the female builds a down-lined hollow in the ground among low plants. She lays one egg per day, but does not start her lone incubation until the full clutch of 7–9 eggs is laid. Often the male stays with her to help guard the exposed nest and to accompany her to the water with the chicks. Female pintails are fearless in the defense of their young, attempting to combat much larger predators. But less than half of the chicks survive to breed the following year because of natural predators and hunters.

 *DID YOU KNOW?*

● The northern pintails in Europe migrate as far south as the Sudan and are depicted in ancient Egyptian paintings.

● When foraging for seeds in the water, the pintail may accidentally ingest spent lead shots from hunters; many pintail succumb to lead poisoning each year.

PROFILE   NORTHERN PINTAIL

The pintail's slim neck and streamlined body make an elegant profile; when swimming or flying, its distinctive tail feathers are easily identified.

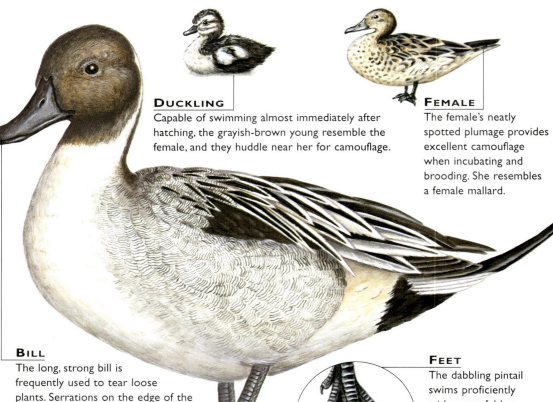

DUCKLING Capable of swimming almost immediately after hatching, the grayish-brown young resemble the female, and they huddle near her for camouflage.

FEMALE The female's neatly spotted plumage provides excellent camouflage when incubating and brooding. She resembles a female mallard.

BILL The long, strong bill is frequently used to tear loose plants. Serrations on the edge of the bill help trap food inside when dabbling.

FEET The dabbling pintail swims proficiently with powerful legs and webbed feet. Since the feet are set farther forward than on diving ducks, the pintail also walks well on land.

VITAL STATISTICS

WEIGHT	1–3 lbs.; male averages 2.3 lbs., female averages 1.9 lbs.
LENGTH	Male 20–29"; female 17–25"
SEXUAL MATURITY	1 year
BREEDING SEASON	April–July
NUMBER OF EGGS	7–9
INCUBATION PERIOD	22–24 days
FLEDGING PERIOD	5–7 weeks
BREEDING INTERVAL	1 year
TYPICAL DIET	Grasses, seeds, tubers, tadpoles, insects and mollusks
LIFESPAN	Up to 27 years

RELATED SPECIES

● The pintail is 1 of 36 species of dabbling, or surface-feeding, ducks in the genus *Anas*, which also includes the mallard, *A. platyrhynchos*, teals and shovelers. Geese and swans join dabbling and diving ducks in the family *Anatidae*; all are considered true waterfowl. This family is 1 of 2 families comprising about 40 genera and over 140 species in the order *Anseriformes*.

CREATURE COMPARISONS

At 23" and 2 lbs., the long-tailed duck (*Clangula hyemalis*) is slightly smaller than the pintail. Only the male has the extended tail feathers. But both sexes have distinct summer, eclipse and winter plumages, which are among the most complex series of plumage patterns of any bird. Unlike the dabbling pintail, the long-tailed duck is a diver. It forages for food in deeper water than any other duck, diving up to 50' or more. The most numerous of all arctic breeding ducks, the long-tailed duck swallows shellfish whole; the shells are ground in its gizzard.

Northern pintail

Long-tailed duck

Northern Wheatear

• ORDER •	• FAMILY •	• GENUS & SPECIES •
Passeriformes	Turdidae	*Oenanthe oenanthe*

KEY FEATURES

● A ground-loving bird of moorland, alpine and open rocky areas, it sings from vantage points while remaining alert for food and predators

● Male performs a feather-fluffing, dancing display when intruders invade its territory

● Migrates long distances between breeding grounds and wintering areas

WHERE IN THE WORLD?

Breeds from Alaska, Canada and Greenland to Europe, the Middle East and northern Asia, including most of Siberia; all but a small population in Iraq overwinter in sub-Saharan Africa

LIFECYCLE

One of the first migratory birds to appear at its breeding grounds, the northern wheatear brings life, color and sound to bleak hills and alpine areas emerging from a long winter.

HABITAT

The northern wheatear breeds in temperate and subarctic regions. Almost all of these birds winter in hot, dry areas of tropical Africa. During the breeding season, it is often found on rocky slopes, scree and alpine meadows at altitudes of up to 10,000'.

Avoiding forests and woodlands, the wheatear prefers open country, and typically breeds in exposed areas rich in insect life, such as hillsides, stony slopes and walled fields where it finds plenty of nesting sites and vantage points. The wheatear is also at home on rocky coasts, bogs and Arctic tundra.

In its winter quarters in sub-Saharan Africa, the wheatear remains a lover of open ground, inhabiting short-grass savannah, farmland, barren rocky hills and plains recently cleared by fire.

▲ **WINDSWEPT ISOLATION**
Breeding birds favor remote, open regions.

? DID YOU KNOW?

● Nearly the entire population of wheatears winters south of the Sahara.

● During migrations, wheatears often cross vast tracts of ocean. Birds from Greenland may take over 30 hours to make the 1,450-mile crossing to Africa.

● The wheatear's name comes from the Anglo-Saxon words *hwit* and *oers*, which mean "white" and "rump."

FOOD & FEEDING

The wheatear feeds mainly on insects, such as beetles, flies and other invertebrates, including snails, slugs and spiders. The bird pounds large insects, such as grasshoppers, against the ground to break off inedible legs and wings. In autumn, when food becomes scarce, the wheatear supplements its diet with berries.

The bird uses two hunting strategies to locate and capture prey. Commonly, it perches on a low rock or bush, watching for any movement on the ground. It then darts down, sometimes hovering above its quarry, before pouncing.

The second method involves hopping over open ground before pausing and watching for prey. If nothing stirs, the bird repeats the process until successful.

◀ **INSECT ADDICT**
While eating, the wheatear looks for more.

HILLSIDE HUNTER

① A fine view…
Perched on a rock, a wheatear scans the bare ground below for signs of prey. Its sharp eyesight can detect the smallest of movements.

② A false move…
As a large wolf spider emerges from hiding, the wheatear swoops down instantly for a closer look.

③ Unequal race…
The spider scuttles for cover, but the bird easily outpaces it and catches it in its slim bill. Large prey offers a highly nutritious snack.

④ Dining at leisure
The bird carries its spoils back to its vantage point. Here, it can safely eat while keeping a close watch for predators.

BREEDING

▲ **Safety coloring**
Brown spotted plumage helps hide the fledgling.

Living in open areas devoid of trees, the northern wheatear must build its nest in crevices in stone walls, rocky ground or in abandoned burrows in order to conceal it from predators, such as weasels and rats.

The nest foundation is an untidy mass of dried stems, rootlets and grass, occasionally decorated with a large feather. This holds a cup-shaped nest constructed of tightly woven grass stems, leaves, moss and lichen.

The pale blue eggs are incubated mainly by the female. Both parents feed the hatchlings, which grow swiftly. After fledging it is important that the young birds put on weight quickly, building themselves up for the long migration south.

▶ **Feeding the family**
Parent birds must hunt continually to feed young.

BEHAVIOR

Being a mostly solitary bird of open country, the northern wheatear is shy and wary — constantly alert to the threat of attack by birds of prey, such as falcons. When alarmed, it flies swiftly for cover uttering a harsh, penetrating *chack-chack* call to warn its mate of the danger.

The wheatear is territorial and returns to the same breeding site each year. If both partners survive the arduous migration, they are likely to pair up once again. Should only one partner survive, he or she waits for a new mate to arrive.

The male defends his territory from intruders with a vigorous, warbling call accompanied by a dancing display. With fluttering wings he leaps rapidly and erratically from side to side, ruffling his breast feathers and flicking his tail. He may even fly over his rival — appearing as a swirling mass of feathers. The rival may attempt to outdo the territory owner with an energetic and vocal display of his own.

▲ **On the stage**
The male sings from a vantage point.

 CONSERVATION

The northern wheatear remains common in its range, but populations in Europe have declined in recent years. This is partly due to the increased use of pesticides, leaving little food for the bird during its breeding season. Forestation of upland areas also reduces the bird's favorite habitats. Like most migrants, many wheatears are targeted by "sport" hunters — particularly in Mediterranean countries — who shoot down songbirds and birds of prey as well as gamebirds.

Profile   Northern Wheatear

Jaunty and restless, the wheatear flits, hops or runs from perch to perch, flicking its wings and flaunting its strongly marked tail if threatened.

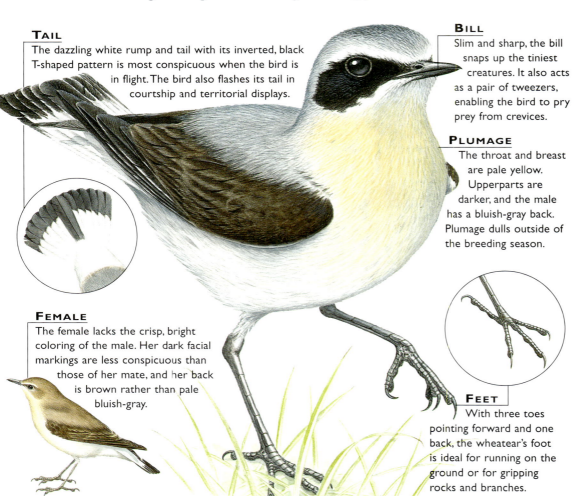

TAIL
The dazzling white rump and tail with its inverted, black T-shaped pattern is most conspicuous when the bird is in flight. The bird also flashes its tail in courtship and territorial displays.

BILL
Slim and sharp, the bill snaps up the tiniest creatures. It also acts as a pair of tweezers, enabling the bird to pry prey from crevices.

PLUMAGE
The throat and breast are pale yellow. Upperparts are darker, and the male has a bluish-gray back. Plumage dulls outside of the breeding season.

FEMALE
The female lacks the crisp, bright coloring of the male. Her dark facial markings are less conspicuous than those of her mate, and her back is brown rather than pale bluish-gray.

FEET
With three toes pointing forward and one back, the wheatear's foot is ideal for running on the ground or for gripping rocks and branches.

Creature Comparisons

Whereas the northern wheatear is a bird of open country, the green cochoa (*Cochoa viridis*), a closely related member of the thrush family, inhabits mountain forests. Found at altitudes of 10,000–16,500' in the Himalayas and other mountains of southeastern Asia, the green cochoa hunts in deep undergrowth for insects. It forages in a similar manner to the wheatear, but its more powerful bill enables it to tackle larger prey. The green cochoa's nest is more refined than that of the wheatear, constructed within tree foliage from mosses, leaves and plant fibers. A far larger bird than the wheatear, the male green cochoa's head is bluish-violet, while his wings and tail vary from gray to powder-blue.

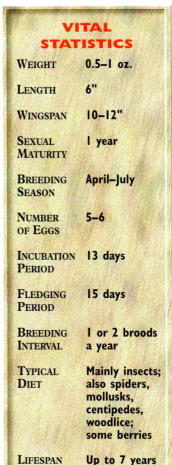

Vital Statistics

Weight	0.5–1 oz.
Length	6"
Wingspan	10–12"
Sexual Maturity	1 year
Breeding Season	April–July
Number of Eggs	5–6
Incubation Period	13 days
Fledging Period	15 days
Breeding Interval	1 or 2 broods a year
Typical Diet	Mainly insects; also spiders, mollusks, centipedes, woodlice; some berries
Lifespan	Up to 7 years

Related Species

● The wheatear belongs to the genus *Oenanthe*, which also contains the desert wheatear, *O. deserti* (below). The genus is part of the family *Turdidae*, which is one of the largest families in the order *Passeriformes*.

OILBIRD

· ORDER ·
Caprimulgiformes

· FAMILY ·
Steatornithidae

· GENUS & SPECIES ·
Steatornis caripensis

KEY FEATURES

- The world's only nocturnal, fruit-eating bird
- Spends its days in a pitch-black cave, only coming out after dusk to feed
- Emits high-pitched clicking sounds like a bat to navigate through the utter blackness of its cave
- One of the few birds with a keen sense of smell

WHERE IN THE WORLD?

Found in northern and central South America from western Guyana through Venezuela and Colombia to Ecuador, Peru and northwestern Bolivia; also found in Panama and Trinidad

LIFECYCLE

The oilbird is a creature of darkness, spending over half of its life in the ink-black depths of a mountain cave, emerging after dusk to feed in the lightless tropical forest.

HABITAT

▲ **CAVE BIRD**
Caves are a safe daytime haven from predators.

By night, the oilbird inhabits the tropical forest of Central and South America. As a fruit eater, it favors mature tropical forest that provides it with a year-round supply of food, and it has a particular fondness for laurel, palm and incense trees.

By day, most oilbirds exploit a very different habitat: deep caves found in the mountains of South America and Panama and on the coast of Trinidad. The oilbird may nest over 2,000' from the cave entrance — the only source of light. Most oilbirds roost in caves, but some populations in Venezuela also roost in palm trees.

DID YOU KNOW?

- The Spanish name for the oilbird is *guácharo*, which means "the one who wails."
- The oilbird is the only non-insectivorous member of the order *Caprimulgiformes*.
- The oilbird is so-called because its nestlings are used by locals as an oil source.
- The oilbird's echolocation can only detect objects larger than 8" wide.

FOOD & FEEDING

In the tropics, night falls quickly and as the shadows lengthen, oilbirds stream out of their roosting caves like a host of bats. As they emerge, they break into small flocks to begin their night-long search for food.

The oilbird has a well-developed sense of smell; scent plays a key role in its search for aromatic fruit. Hovering on its long wings, the bird plucks oily fruit of palm and laurel trees with its hooked bill. Lacking a *crop* (a saclike part of the esophagus), it carries food in its stomach to be digested during the day in its cave.

▲ **EATING OUT**
The oilbird emerges from its cave at night to feed.

NIGHT BIRDS

❶ Colonial life...
The oilbird roosts colonially in deep caves, taking wing at night to search the neighboring tropical forests for juicy fruit.

❷ Steering by echo...
In the cave's darkness, the oilbird produces clicking sounds that bounce off the walls and give it a picture of its surroundings.

❸ Led by its nose...
Travelling up to 15 miles over the forest, the oilbird uses its keen sense of smell to locate a tree laden with ripe fruit.

❹ Struck oil
Swooping down, the oilbird hovers skillfully as it eats its fill, picking off the oily fruit of a palm tree with its sharply hooked bill.

BEHAVIOR

The oilbird is highly social; a large cave may have hundreds of roosting birds. Despite the numbers, the cave is often eerily quiet by day. If a roost is disturbed by an intruder, the silence is shattered by loud, shrieking, alarm calls.

The oilbird navigates its cave using *echolocation*, a kind of sonar similar to, but cruder than, a bat's. Producing clicking sounds as it flies, the returning echoes build up a picture of the surroundings. These sonar clicks are of a lower frequency than those of bats and are audible to humans.

When flying at night, the oilbird relies entirely on its large, light-sensitive eyes to find its way. Its flight above the trees is light, swift and undulating.

▶ **HEAD TO HEAD**
Oilbirds are social birds that pair for life. They often preen each other's head to strengthen bonds.

▶ **THE EYES HAVE IT**
Outside its cave, the oilbird relies on keen vision.

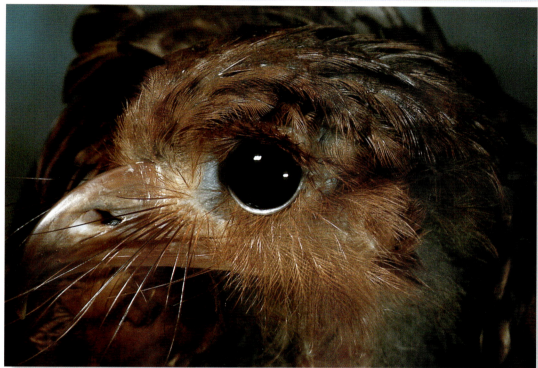

 ## BREEDING

▼ **NATURAL BIRTHPLACE**
Nests are made from regurgitated food and droppings.

In the seasonless tropics, oilbirds breed year-round. Pairs mate for life and use the same nest every year to raise their young. Two to four eggs are laid, and both parents incubate them for 32–35 days.

Young are fed on a diet of fruit regurgitated by their parents three to four times a night. This diet quickly makes the chicks fat, so that by 70 days, when feathers first appear, they weigh about half as much as their parents. As they master the art of flying in the darkness of their cave, the young birds steadily lose weight. Young are fledged by the time they are 90–120 days old.

 **CONSERVATION**

Not yet endangered, the oilbird is in decline due to ongoing destruction of its tropical forest habitat — and this poses a serious threat to the bird's future. Although the tradition of collecting nestlings for oil production is less common today, it persists in remote areas where protection is difficult to enforce.

Profile Oilbird

Supertuned senses of hearing, sight and smell help the oilbird find aromatic fruit in the forest at night and locate its nest deep in a cave.

Ears
Acute hearing analyzes the pulses of sound that it sends out when navigating the pitch blackness of its roosting cave.

Eyes
Large, light-sensitive eyes give good vision in low light, enabling the oilbird to fly safely through the forest at night in search of food.

Bill
The eaglelike hooked upper mandible is adapted to plucking fruit from trees rather than tearing flesh. The stiff feathers around the bill provide the oilbird with a sense of touch.

Wings
The oilbird's long wings help it hover when it is picking fruit off trees.

Nostrils
Unlike most other birds, the oilbird has large nasal cavities covered with scent-detecting mucous membranes that give it a well-developed sense of smell. The oilbird relies entirely on scent to locate fruit in the dark.

VITAL STATISTICS

Weight	13–17 oz.
Length	1.5'
Wingspan	3–4'
Sexual Maturity	1 year
Breeding Season	Year-round; a peak from December–June
Number of Eggs	2–4
Incubation Period	32–35 days
Fledging Period	90–120 days
Breeding Interval	1 year
Typical Diet	Fruits, mainly from palm and laurel trees
Lifespan	12–15 years

RELATED SPECIES
● The oilbird is the only member of the family *Steatornithidae*, but shares the order *Caprimulgiformes* with nocturnal birds, including frogmouths of Southeast Asia and Australia, potoos of the West Indies and South and Central America, frogmouths of Australia and Papua New Guinea and the several nightjar species of Eurasia and North America.

CREATURE COMPARISONS

Despite being half the size of the oilbird, the Eurasian nightjar, *Caprimulgus europaeus*, shares the same body shape and long, sickle-shaped wings. Both feed at night. The oilbird spends its days in a deep cave; the nightjar relies on its camouflage and perches, branchlike, along a tree limb. The hook-billed oilbird feeds on fruit, but the nightjar opens its tiny bill wide to catch moths and other insects on the wing, aided by stiff feathers that help funnel prey into its mouth. Both birds are strong, agile fliers, but while the oilbird can find food in its tropical habitat year-round, the nightjar flies south to Africa when the weather cools and insect prey disappear.

Eurasian nightjar   Oilbird

OSPREY

• **ORDER** •
Accipitriformes

• **FAMILY** •
Pandionidae

• **GENUS & SPECIES** •
Pandion haliaetus

KEY FEATURES

- Remarkably powerful for its size, it can seize and fly off with prey almost as heavy as itself
- Spiny-soled feet, tipped with razor-sharp talons, enable it to grip the most slippery of prey
- Most widespread of all birds of prey, occurring on every continent except Antarctica

WHERE IN THE WORLD?

Breeds throughout much of Europe and Asia, North America, Southeast Asia and Australasia; winters in South America, Africa and India

LIFECYCLE

The osprey, also known as the fish hawk, is an adaptable, eagle-sized bird that preys almost exclusively on fish in nearly every fresh- and saltwater habitat in the world.

HABITAT

▲ **WATERSIDE BIRD**
The osprey exploits any habitat where it can fish.

The osprey's range is limited only by the availability of its specialized diet. It can live wherever it has year-round access to clear, unpolluted waters — salty or fresh — that contain plentiful supplies of small- to medium-sized surface-feeding fish.

Favored saltwater habitats include rocky, sandy or forested coastlines, remote oceanic islands, salt marshes and tropical mangrove swamps. Inland, it's found near lakes, rivers and marshes. In all habitats, the osprey seeks tall trees to perch and nest in, but will also use electricity poles, harbor buoys, bridges and abandoned buildings.

FOOD & HUNTING

BEHAVIOR

The osprey flourishes in a wide range of climates, but northern populations migrate south in winter as waters become cold or frozen, driving fish to greater depths. Individual ospreys migrate singly, but often in the company of other birds of prey, such as red-tailed hawks and kestrels. At this time, loose colonies of as many as 30 birds sometimes gather on estuaries or along rich stretches of coastline.

▲ **SITTING PRETTY**
Between dives, the osprey dries, preening its feathers.

The osprey devotes much time to keeping its plumage in good condition, removing any sticky fish scales and reapplying oil from the preen gland above its tail to keep its feathers waterproofed.

CONSERVATION

Although still widespread and numerous, many populations of osprey are declining due to pollution and overfishing. A large number of ospreys are also shot each year on migration.

678 Osprey

When hunting, the osprey patrols up and down a stretch of water, flying with deep, relaxed wingbeats as it searches for a fish below. It may hover clumsily over the same spot before closing its wings and plunging for the kill.

It feeds on fish, including trout, salmon and carp from fish farms and stocked lakes, but also takes carrion, small birds and mammals.

◀ **SHARP IMPLEMENT**
Once a fish is caught, the osprey returns to a perch and rips it apart with its bill.

DID YOU KNOW?

● Sparrows occasionally make their homes within the sprawling osprey's nest.

● The osprey has such large feet that sometimes it can snatch two small fish from the water in a single dive.

● Less than half of an osprey's dives capture a fish. Strong winds and murky water can affect the accuracy of its strike and reduce its success.

PERFECT TIMING

❶ Search...
Flapping slowly 65–100' above a lake, its wings held in a M-shape, an osprey searches the water for prey.

❷ Stoop...
Spotting a fish below the water's surface, the bird draws in its wings and plunges at a 45° angle.

❸ Snatch...
Judging its final attack with superb accuracy, the osprey dives feetfirst to snatch the fish from the water.

❹ Success
Struggling into the air with its load, the bird shakes water from its wings and flies to its perch.

BREEDING

Breeding seasons depend on the latitude of the osprey's nest site: January in subtropical regions; May and June in northern areas.

Ospreys mate for life, each pair refurbishing the same substantial nest year after year. The female lays three eggs and incubates them alone; the male relieves her for short periods while she hunts. Once the first chick hatches, the male fishes for the family, bringing up to seven fish to the nest each day. Before delivering them, he eats their heads so the fish can't damage eggs or young by floundering around. After six weeks, the female helps the male feed the chicks until they fledge.

▲ **DRAMATIC DISPLAYS**
A pair mates on a perch after a showy courtship display of dramatic swoops and wing displays.

▼ **GROWTH SPURT**
Due to its protein-rich fish diet, the young osprey fledges at 7–8 weeks.

Profile Osprey

Sharp eyes, narrow wings and dexterous, powerfully taloned feet equip the osprey to find, capture and kill the fish on which it feeds.

FEMALE
Larger than the male, the female also has a more prominent band of brown feathers across the breast.

BILL
Sharply hooked for piercing and ripping open the tough and scaly skin of fish.

PLUMAGE
Denser and more oily than the plumage of most birds of prey, the osprey's feathers shrug off water with ease.

WINGS
For a bird of prey with such a wide wingspan, the wings are unusually narrow. But wider wings would trap too much water in the feathers, making it harder for the osprey to get airborne after a dive to catch prey.

FEET
Outer toe on each foot rotates backward to help the bird grasp prey firmly. Spiny soles help hook the fish to the feet.

LEGS
Long legs increase the bird's reach, enabling it to catch fish without submerging its body in the water.

VITAL STATISTICS

WEIGHT	Male 3 lbs.; female 3.5 lbs.
LENGTH	About 2' for both sexes
WINGSPAN	5'
SEXUAL MATURITY	3 years
BREEDING SEASON	Varies with range
NUMBER OF EGGS	2–5; usually 3
INCUBATION PERIOD	34–40 days
FLEDGING PERIOD	49–57 days
BREEDING INTERVAL	1 year
TYPICAL DIET	Almost entirely freshwater and marine fish
LIFESPAN	5–15 years

RELATED SPECIES

● The unique structure of the osprey's feet led to it being classified in its own family: *Pandionidae*. In the 220 species of *Accipitriformes*, there are several fish eaters, including the African fish eagle, *Haliaeetus vocifer* (below).

 CREATURE COMPARISONS

Larger than the osprey, with a wingspan of 6–8' and up to 7 lbs., the African fish eagle (*Haliaeetus vocifer*) is also an accomplished predator. Having spotted a flash of silver from its waterside perch, the eagle flies over the water and deftly snatches its victim with a single foot, barely missing a wingbeat as it does. Only when catching very large fish does the eagle partly submerge itself. Where its range overlaps with the osprey's, the eagle isn't above a little aerial piracy and will often bully its smaller relative into giving up its catch.

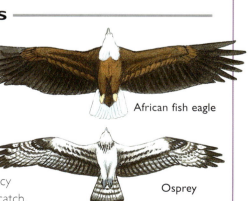

African fish eagle

Osprey

Ostrich

• ORDER •	• FAMILY •	• GENUS & SPECIES •
Struthioniformes	*Struthionidae*	*Struthio camelus*

KEY FEATURES

● Lives on open plains where its "watchtower" vision helps it spot predators; its group-based lifestyle also gives the protection essential for exposed plains dwellers

● Gapes its bill and spreads its wings to dissipate body heat

● Gleans all of its water needs from plant food — except during the driest months

WHERE IN THE WORLD?

In a strip across Africa between 10° and 20° north of the equator — East Africa south down the Rift Valley to Tanzania; also inhabits South Africa, Namibia, southern Angola and Botswana

LIFECYCLE

The ostrich may be flightless, but it can run extremely fast — which is how this huge, nomadic bird escapes from predators as it roams the open grasslands in search of food.

HABITAT

▲ **WALK TALL**
The ostrich's great height suits it to plains life.

The ostrich lives on the short-grass savannahs of Africa and in semidesert regions. It wanders far in search of food, moving to desert areas when vegetation springs up after seasonal rains.

The ostrich avoids long grass, where predators may lie in wait and shuns thick woodland for the same reason. It favors undulating land, which allows it to feed inconspicuously and where it can run away at the first sign of danger. Any areas where humans have altered the landscape in any way are avoided.

CONSERVATION

Traditionally, the ostrich was hunted for meat, eggs and hide, as well as feathers. A more recent demand for its meat has prompted growth in commercial farming; this has relieved hunting pressure on wild birds, whose status is stable. Today, the ostrich has to compete with domestic grazing stock on its native grasslands, but it prospers in reserves. Attempts are being made to reestablish it in its former range.

BREEDING

The male performs a courtship dance in front of a female and prepares several nests for her. After mating, the female selects a nest and lays her first egg, which is 6" long. As the first to be mated, she becomes the dominant female of the group. She lays an egg every two days, to a total of up to 11 eggs.

The male also mates with several other females. They all lay clutches in the same nest; there may be 20 to 30 eggs when the dominant female is ready to begin her 42- to 46-day incubation. If there are too many eggs to brood, the female rolls a few from the nest. The whole clutch may hatch in one or two days. Chicks are first guarded by both the male and female but later join other broods to form a crèche, reaching full size in 18 months.

COURTING COUPLES

❶ Let courtship begin...
The male approaches the female with straight neck, erect, fluffed tail and drooping wings.

❷ Swooning and swaying...
When close, he drops down with wings outspread. He sways his neck, while raising his wings alternately.

BEHAVIOR

The ostrich normally lives in small groups (five to ten birds), although larger flocks assemble around water in the dry season or where food is abundant.

▶ **ALL TOGETHER NOW...**
Large flocks gather around water in the dry season, but spend most of their time in smaller family groups.

Groups stride across the short grass plains, frequently picking food from the ground. When not feeding, the ostrich spends its time resting, dust bathing and preening.

With its excellent hearing and height, the ostrich is often the first animal on the plains to spot a predator; its fleeing flocks often alert other animals to danger. The ostrich can reach speeds of 39 mph and outpaces most enemies. Rarely, it kicks powerfully in self-defense.

❸ Back beat…
As the display gets frenzied, his neck writhes violently and his head thumps on his broad back.

❹ Check mate
When the female is ready, they walk together. She lies down and the male mounts her.

FOOD & FEEDING

The ostrich feeds in a small group, stooping for plant material (mainly herbs, grasses, seeds and flowers) and occasionally insects and small lizards. It swallows several items together, which can be seen travelling down the neck as a small lump or *bolus*.

In dry areas, the ostrich browses on succulent plants, which may provide all its water needs, although it will drink regularly when water is available. It also swallows small stones and grit to help break down and digest plant matter. In captivity, an ostrich may swallow all manner of strange objects as substitutes for this digestive grit: one zoo bird met an untimely end after attempting to digest a 3' length of rope.

▼ **Drink up!** Ostriches are especially on guard for predators at waterholes.

 *DID YOU KNOW?*

● When threatened, an incubating ostrich will lay her neck and head flat on the ground. This may have given rise to the legend that a threatened ostrich buries its head in the sand.

● A man weighing 250 lbs. can stand on an ostrich egg without breaking it.

● Egyptian vultures have been seen cracking open ostrich eggs by dropping large stones on them.

PROFILE OSTRICH

The ostrich's powerful legs equip it to roam the open plains with ease, while other special features help it cope with the relentless heat and dust.

BILL
The broad gape of the bill helps the ostrich dissipate excess body heat during the hottest hours of the day.

WINGS
The ostrich is flightless, but uses its wings for display and heat regulation.

LEGS
Legs are muscular and huge for running at high speeds and delivering defensive kicks. Thighs are naked to improve heat dispersion while it is running.

EYES
Huge eyes scan the savannah for danger. Long lashes protect eyes from dust storms or debris kicked up while running.

STERNUM
The flattened sternum (breastbone) lacks a keel, which serves as an anchor for flight muscles in other bird species.

FEET
The ostrich has only two toes, a feature unique among birds. The large toe has a flat nail for fast running, and the second toe aids balance.

VITAL STATISTICS

WEIGHT	Female about 249 lbs.; male up to 339 lbs.
HEIGHT	6–8'
SEXUAL MATURITY	Female 2 years; male 3–4 years
MATING SEASON	Variable, often dependent on local climatic conditions
NUMBER OF EGGS	Each female lays 5–11 eggs
INCUBATION PERIOD	42–46 days
BREEDING INTERVAL	Unknown
TYPICAL DIET	Mostly plant material; some invertebrates
LIFESPAN	Up to 40 years

RELATED SPECIES
● The ostrich is the only member of the family *Struthionidae*. Other flightless birds around the world are the rheas of South America, the kiwis of New Zealand and the cassowaries (below) and emus of Australia.

CREATURE COMPARISONS

At up to 8' tall, the male ostrich is the largest living bird. Its flightless relative, the Australian emu, stands almost 6' tall. Forest-dwelling birds called moas, which were even taller and heavier than ostriches, roamed New Zealand as recently as 300 years ago. Of the 12 moa species, the largest was the giant moa, which stood 10' tall or more.

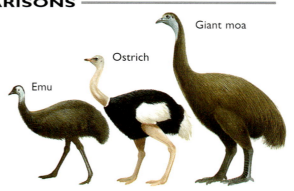

Ovenbird

• ORDER •	• FAMILY •	• GENUS & SPECIES •
Passeriformes	Parulidae	Seiurus aurocapillus

KEY FEATURES

• Easily identified by its unusual walking gait and carriage; its tail tilts upward and wings droop slightly

• Takes its name from the shape of its nest, which looks like an old-fashioned Dutch oven

• To lure predators away from its nest, the female ovenbird will flutter away with wings and tail dragging as if mortally wounded

WHERE IN THE WORLD?

Found across North America, north to Newfoundland and Alaska; west to the Rockies; south to the West Indies and Bahamas; abundant in Florida, South Carolina and Louisiana

Ovenbird 685

LIFECYCLE

The ovenbird stays close to the forest floor where it finds food and keeps its unusual oven-shaped nest well hidden among the beds of fallen leaves.

HABITAT

The ovenbird frequents forests where its natural camouflage blends in best. Deciduous and mixed forests both suit the ovenbird, which breeds in the Canadian northwoods and in the northeastern U.S. During fall and winter, the ovenbird migrates to Mexico, Central America and as far south as Venezuela, but it is frequently spotted in winter in the far southern reaches of the U.S.

▲ **SERENE SCENE**
The ovenbird prefers forests with lots of shade and leaf litter.

FOOD & FEEDING

The ovenbird forages most of the day on the forest floor. It deliberately strides, with its high-stepping gait, across leaf litter and logs as it searches for tidbits. Throughout the day, the ovenbird is safely camouflaged in the cover of fallen leaves. The ovenbird turns over dead leaves to glean snails, earthworms, spiders and, on occasion, uncovers a small lizard or frog. In the fall and winter, it relies on fruit and other vegetation, since insects are scarce. The ovenbird's bill opens seeds with ease. The ovenbird also consumes grit, used in the bird's gizzard to help grind up food. The bird picks very little prey from live vegetation, but it occasionally will alight in trees to feast on tree snails or hatches of budworms found in spruce trees. The normally reclusive ovenbirds have even been seen at Everglades National Park feeding on table scraps left over from tourists.

▼ **GOBS OF GRUBS**
The inquisitive ovenbird leaves no leaf unturned and uncovers a mouthful of larval grubs.

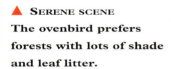
DID YOU KNOW?

● Female ovenbirds are so reluctant to leave their ground nest containing young that they end up getting stepped on rather than fleeing.

● During a 48-hour period, a noisy ovenbird sang 215 times, more than once every 13 minutes, day and night.

● The genus name, *Seirus*, comes from the Greek "seio," to move, and "oura," meaning tail. The species name, *aurocapillus*, comes from the Latin words, "*aurum*" and "*capillus*," meaning gold hair, referring to the ovenbird's golden-orange crown.

CONSERVATION

Ovenbird populations have been in decline for more than two decades, mainly due to habitat loss. The bird can be almost absent in tracts of 250–2,000 acres. Radio towers and skyscrapers take a toll on migrating flocks.

BEHAVIOR

The call of the ovenbird, *tea-cher, tea-cher, tea-cher,* is a familiar sound during spring in the forests of the Northern U.S. and Canada. It also emits a warning *cheep* if it detects something approaching its nest. The ovenbird has a distinctive walk. With its tail up, wings drooped and head bobbing, it is easily distinguished from other warblers as it crosses the forest floor. Its coloring provides such effective camouflage that it can wait until the last moment to react to a threat. It migrates in flocks but is solitary outside the breeding season; some are permanent residents in the south.

▶ **SHADES OF BROWN**
The bird's plumage blends with the forest floor.

BREEDING

In spring, the male ovenbird arrives at its breeding grounds about two weeks earlier than the female. Once she arrives, he puts on an impressive aerial display, on occasion flying 20' above her, circling and then making a downward flight. He then hops, struts and bows his head toward the female. If she accepts his attentions, they mate. The nest is made of leaves, stems, bits of moss and bark. Hidden on the ground, it's shaped like an old-fashioned Dutch oven (hence the bird's name) with a shallow dome over the top and an opening on the side. When the nesting female is approached she occasionally will act if she is hurt, fluttering away from the nest to attract attention to herself. The ovenbird lays 3–6 creamy-colored eggs with brown spots. It must beware of the cowbird, which pushes aside an ovenbird egg and lays one of its own as a replacement. When this happens, the ovenbird raises the chick as her own. The hungry nestlings are fed insects dozens of times per day and fledge in 8–11 days.

HOME SWEET DOME

❶ Call of the wild...
An ovenbird male sings its familiar song from the safety of the forest floor.

❷ Under construction...
A female ovenbird uses twigs and leaves to build her oven-shaped nest in the deep beds of leaf litter.

❸ Over the top...
The female weaves the canopy covering the top of the nest, concealing it from predators.

❹ Finishing touches
With the dome finished, the nest's lining is padded with plant stalks and even horse hair.

PROFILE OVENBIRD

The ovenbird's distinctive song echoes through the woodlands while its natural camouflage keeps it well hidden on the forest floor.

BILL
The short, strong and conical beak is a very efficient tool for opening the seeds that make up the bulk of the bird's diet.

FEET
With three toes pointing forward and one pointing backward, the ovenbird's foot is built for perching. Its grip automatically tightens if its weight shifts backward.

JUVENILE
The juvenile's olive-buff head will begin to show the adult's golden-orange hue within its first year. The bold, black spotting will align into neat rows as it matures.

IN FLIGHT
The ovenbird spreads its wings and tail and flies 10–60' above the treetops to sing its song during the breeding season.

VITAL STATISTICS

WEIGHT	0.5–1 oz.
LENGTH	5.75"
WINGSPAN	9–11"
SEXUAL MATURITY	1 year
BREEDING SEASON	May–July
NUMBER OF EGGS	3–6, usually 4–5
INCUBATION PERIOD	11–14 days
FLEDGING PERIOD	8–11 days
BREEDING INTERVAL	1 year
TYPICAL DIET	Seeds, snails, earthworms, insects, spiders, fruit, lizards and frogs
LIFESPAN	8 years

CREATURE COMPARISONS

Unlike most warblers, including the ovenbird, the Prothonotary warbler (*Protonotaria citrea*) is a cavity nester that makes its home in tree holes or other similar cavities. The Prothonotary warbler is also known as the golden swamp warbler because of its brilliant coloration and preference for damp, dark lowland woods as nesting sites. This warbler also favors wooded swamps, flooded bottomland forests and areas surrounding slow-moving water, in sharp contrast to the ovenbird, which nests mainly in dry, grassy areas.

Ovenbird

Prothonotary warbler

RELATED SPECIES

● The ovenbird is one of only three species in the genus *Seiurus*. The northern waterthrush, *S. noveboracensis*, has a similar gait to the ovenbird, but prefers streams and pools. The Louisiana waterthrush, *S. motacilla*, also tends to live near streams. The order *Passeriformes* includes 46 families of perching birds, such as broadbills, larks, warblers and honeyeaters.

Oxpeckers

- **ORDER** · Passeriformes
- **FAMILY** · Sturnidae
- **GENUS & SPECIES** · *Buphagus erythrorhynchus, B. africanus*

KEY FEATURES

- Make a living by feeding on the parasitic insects that infest big game and cattle

- Hitch a ride on their hosts, snapping up prey with a scissorlike bill

- Tolerated by their hosts because they perform a valuable service

- Nest cooperatively, with each breeding pair being supported by up

WHERE IN THE WORLD?

Found from Sudan and Ethiopia in East Africa, across to the western coast of the continent, and south through Kenya, Uganda and Tanzania to South Africa

LIFECYCLE

Despite their name, oxpeckers do not limit their cleaning services to oxen. They can be found on many African mammals, nipping away at the parasites that infest their skins.

HABITAT

Oxpeckers are birds of the African savannah, where open, grassy plains are interspersed with light woodland and scattered shrubs. Their dependence on game animals means that they are common only in protected parks and reserves, where large herds live.

Because they nest in tree cavities, oxpeckers tend to be found in slightly wetter areas of savannah, where larger trees occur. As well as obtaining food from their host, oxpeckers drink from the same waterhole and simply fly down from the animal to the water, and back again once they have quenched their thirst.

▲ **MIDDAY AT THE OASIS**
The oxpecker frequents waterholes on savannah.

CONSERVATION

Where game herds are numerous, the oxpecker is not under threat. It is declining in areas where farmers remove ticks from livestock with chemicals.

FOOD & FEEDING

BREEDING

The oxpecker is a cooperative breeder. Each pair is assisted by up to four nonbreeding birds, which are usually offspring from previous broods. Breeding pairs mate on the ground or on the back of their mammal host.

The pair normally places its nest high in a tree hole, but it may also nest under the eaves of buildings or in thatched roofs, rock cavities and embankments. The nest is an untidy, cup-shaped pad of grass and straw, lined with animal hair plucked from a host, and set on a foundation of dried dung.

The oxpecker usually lays 2–3 eggs, although clutches of one or even five have been recorded. The pair shares the incubating duties during the day, but only the female incubates at night. After two weeks the eggs hatch, and the chicks emerge thinly covered in down. Both parents care for the young, and helpers assist in feeding the brood. The chicks leave the nest after about four weeks, but are fed by their helpers and parents for three more months.

▶ **FAMILY OUTING**
Dark juveniles may feed with adults.

The key to the oxpecker's feeding technique is its ability to hitch a ride on the backs of large grazing mammals. The oxpecker visits a wide range of hosts, including the rhinoceros, buffalo, giraffe, zebra and hippopotamus, as well as large antelope species.

Ticks and other blood-sucking parasites make up nearly all of the oxpecker's diet; ticks that are bloated with fresh blood provide the most nutrition. The oxpecker occasionally nibbles around the edges of old wounds, feeding on bits of loose skin, scar tissue and freshly flowing blood. It sometimes hunts on the wing, but is much less adept at this method of feeding.

? DID YOU KNOW?

- Oxpeckers may act as sentinels, giving their host early warning of approaching predators.

- Oxpeckers are also known as tickbirds.

- Elephants, waterbuck and hartebeest do not tolerate oxpeckers.

- Like other members of the starling family, oxpeckers have a range of calls, including shrill whistles and harsh chattering sounds.

- The red-billed oxpecker is absent from the western parts of the genus's range.

A MOVABLE FEAST

① Time to eat...
A flock of hungry oxpeckers heads out over the savannah in search of mammals that are in need of its pest-removing services.

② Mane course...
Clambering about a giraffe's head, an oxpecker carefully probes the coat, looking for ticks and bloodsucking flies.

③ Necking...
Turning its bill sideways, flat against the animal, the oxpecker uses a scissorlike action to cut a tick free, before swallowing it whole.

④ The perfect host
Ticks are found even on the nostrils, ears and eyelids. Buffalo tolerate the oxpecker in these rather sensitive areas.

BEHAVIOR

During the day, the oxpecker travels in small groups, frequently uttering a sharp, hissing call as it flies about between the animals in a herd. At night, the oxpecker returns to a communal roost in a tree or a stand of reeds in a marsh. Sometimes a small party spends the night on its host's back, perhaps to keep warm during cool weather.

The distribution of the two species depends on their hosts. Both the red-billed and yellow-billed species favor less densely furred animals. Where they occur in the same area, the yellow-billed oxpecker seems to have first choice and is usually the species to be seen on sparsely haired mammals.

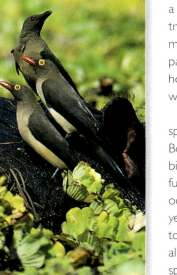

▼ **BACKPACKING** Oxpeckers groom browsers such as kudu.

Oxpeckers 691

PROFILE OXPECKERS

Sharing the powerful feet and stiff, proplike tail of woodpeckers, oxpeckers can keep a secure grip on animals as they walk along.

BILL
The stout bill is flattened on both sides. The oxpecker uses its bill to work through the fur of its host. The yellow-billed oxpecker has a yellow base on its bill.

JUVENILE
A young oxpecker is darker brown than its parents. Its bill is dark olive at first, but gradually takes on adult coloration after four months.

WINGS
Held half-open, the oxpecker's wings help it balance on its mobile perch.

FEET
The short, strong feet have three toes facing forward and one to the rear, each armed with a long, sharp claw.

Species illustrated: Red-billed oxpecker, *Buphagus erythrorhynchus*

TAIL
Long, stiffened tail feathers brace the oxpecker against its hosts' bodies in much the same way as woodpeckers' tails brace them against tree trunks.

VITAL STATISTICS

WEIGHT	1.75 oz.
LENGTH	8–9"
WINGSPAN	12"
SEXUAL MATURITY	1 year
BREEDING SEASON	Wet season: September–February
NUMBER OF EGGS	1–5, usually 2 or 3
INCUBATION PERIOD	12–14 days
FLEDGING PERIOD	26–30 days
TYPICAL DIET	Mainly ticks, bloodsucking flies and lice; some loose skin and blood from open wounds
LIFESPAN	Unknown

RELATED SPECIES

- There are 2 species of oxpecker in the genus *Buphagus*: thered-billed oxpecker, *B. erythrorhynchus*, and the yellow-billed oxpecker, *B. africanus*. They are confined to sub-Saharan Africa, but the red-billed species is less common. Oxpeckers belong to the family Sturnidae, which contains about 105 other species, including the introduced European starling.

CREATURE COMPARISONS

Cattle egret

The cattle egret (*Bubulcus ibis*), a member of the heron family, is not a relative of the oxpecker but shares its savannah habitat. The two birds are often found alongside each other, associating with large grazing animals. However, the egret does not get its food directly off the animals' bodies. Instead, it walks alongside them snapping up grasshoppers, locusts and beetles flushed from the vegetation as the heavy grazers wander over the grassland. Unlike the oxpecker, the egret is an uninvited guest and gives nothing back to the grazers in return for its meals. Standing about 20" tall, the egret is much bigger than the oxpecker. Long legs enable it to walk quickly near feeding herds, and large feet offer stability.

Red-billed oxpecker

Painted Bunting

• ORDER •	• FAMILY •	• GENUS & SPECIES •
Passeriformes	*Emberizidae*	*Passerina ciris*

KEY FEATURES

- Only the male sports the striking blue head and red underparts
- The male's song advertises his presence to adversaries and potential mating partners
- Fiercely territorial, the painted bunting will engage in midair fights to protect its space

WHERE IN THE WORLD?

Found in the southern U.S. and south into Mexico and Cuba; winters from the Gulf of Mexico area to Central America

LIFECYCLE

The beautiful painted bunting rarely emerges from the deep plant cover it prefers; when it is visible, it is often fighting for territory with another bunting.

HABITAT

Within its range in North and Central America, the painted bunting frequents woodland edges where food and water are plentiful. It can also be found in thickets beside streams, gardens and roads. This highly territorial bird values its privacy and prefers to remain close to the protective cover of vegetation.

BEHAVIOR

Practically the only time the normally shy painted bunting is visible is during its particularly fierce territorial battles. Males will protect their territories from intruding males through vicious fights, which can begin in the air and eventually end up on the ground. A bunting may also warn intruders with a threat posture, standing erect with its tail and head held high, forming its distinctive blue head feathers into a small crest. Once a bird is victorious, he sings to advertise his dominance and proclaim the territory as his own. After the breeding season, adult and young birds gather in loose flocks and fly south to warmer climates.

GET OFF MY BRANCH!

① Warning...
The male painted bunting perches in a branch high in a tree, singing his warning call as he holds his head erect with the bill open.

② Serious posturing...
The highly territorial male stands up in a threatening posture, holding both his head and tail high as an intruding male flies nearby.

BREEDING

The male painted bunting arrives north first and establishes a well-defined territory, often occupying the same breeding area as the year before. Males perch at prominent points and sing to establish territories. After chasing away potential suitors, the male sings a courtship song from his perch until a willing female appears. He then dives and chases her through the vegetation with such zeal that the pair sometimes ends up on the ground. After mating, the birds build a cup-shaped nest made of grasses, stalks and leaves in a bush or low tree, or in hanging Spanish moss about 25' off the ground. The male dutifully feeds the female as she incubates the clutch of 3–5 pale-blue eggs for about 12 days. After the young nestlings hatch, both parents feed the chicks until the fledglings leave the nest, about two weeks later. Adults may raise up to four broods each breeding season.

▲ **PRIVATE PERCH**
The shy painted bunting prefers to perch in trees or shrubs well hidden in thickets of vegetation.

◀ **BRANCH BEAUTY**
Bursting with glorious color, a perched female painted bunting decorates a tropical tree branch.

▶ **TWO'S COMPANY...**
Two bunting nestlings wait for their tardy sibling to hatch from its egg.

694 Painted Bunting

③ **Intruder alert...**
The intruding male swoops down and prepares to peck at the male on his perch; the other moves out of the way to avoid the attack.

④ **Fierce battle**
The males twist and turn, feet locked, as they peck at each other with their sharp beaks; they may tumble toward the ground.

CONSERVATION

Although considered common by the National Audubon Society, the bunting is threatened by the loss of its breeding habitat due to the development of swampy thickets and woodland edges. In addition, hundreds of the birds are captured in their tropical wintering grounds each year for the pet trade.

FOOD & FEEDING

Both in the trees and on the ground, the painted bunting feeds on a variety of seeds, which comprise the majority of its diet. The bird's stout, conical bill is well suited for crushing and husking seeds that are found scattered on the woodland floor. It also eats many insects, including caterpillars and flies. The painted bunting will forage most of the day and may also pluck various insects from the trees and surrounding bushes and shrubs to eat or to feed its nestlings.

▼ **DOWN THE HATCH**
A female bunting takes turns feeding its two hungry hatchlings.

DID YOU KNOW?

● According to Native American legend, the Great Spirit created the painted bunting from dabs of many colors left over after all the other birds were created.

● The beauty of the painted bunting has inspired many names. Spanish speakers call it *mariposa*, the butterfly, while in the southern U.S. it is called *nonpareil* (without equal) for its peerless beauty.

Painted Bunting 695

PROFILE PAINTED BUNTING

The painted bunting's electrifying colors are easy to spot as the bird flies in quick bursts, defending its territory.

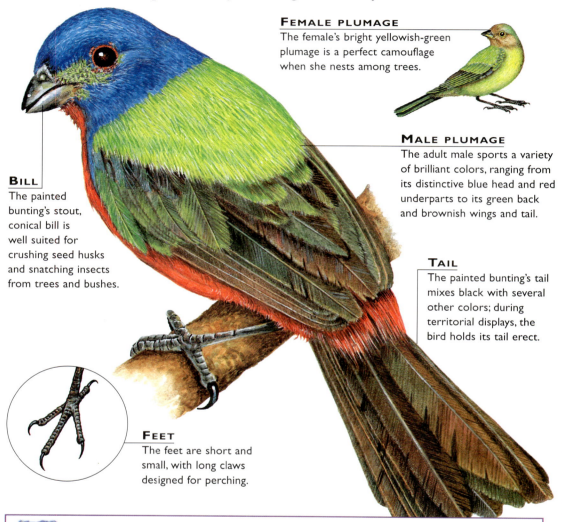

BILL
The painted bunting's stout, conical bill is well suited for crushing seed husks and snatching insects from trees and bushes.

FEMALE PLUMAGE
The female's bright yellowish-green plumage is a perfect camouflage when she nests among trees.

MALE PLUMAGE
The adult male sports a variety of brilliant colors, ranging from its distinctive blue head and red underparts to its green back and brownish wings and tail.

TAIL
The painted bunting's tail mixes black with several other colors; during territorial displays, the bird holds its tail erect.

FEET
The feet are short and small, with long claws designed for perching.

VITAL STATISTICS

WEIGHT	0.5 oz.
LENGTH	5–5.5"
WINGSPAN	9–10"
SEXUAL MATURITY	About 1 year
BREEDING SEASON	March–July
NUMBER OF EGGS	3–5
INCUBATION PERIOD	About 12 days
FLEDGING PERIOD	2 weeks
BREEDING INTERVAL	Up to 4 broods per year
TYPICAL DIET	Seeds, spiders and insects
LIFESPAN	Over 12 years

CREATURE COMPARISONS

The snow bunting (Plectrophenax nivalis) breeds in the colder climates of Iceland, Scandinavia, Scotland, subarctic Asia and North America, while the painted bunting enjoys a much milder climate; it is found in southern North America and Central America. Larger than the painted bunting at 6.25", the snow bunting feeds on a similar diet of seeds and insects. It prefers open, stony country, tundra and mountains, and winters along the coasts. During breeding season, the male snow bunting is mainly pure white except for his black back, tail and primary feathers.

Snow bunting

Painted bunting

RELATED SPECIES

● Among the 6 species in the genus *Passerina* are the indigo bunting, *P. cyanea*, and the lazuli bunting, *P. amoena*. The family *Emberizidae* contains 47 species, including the yellow warbler, *Dendroica petechia*, and the great-tailed grackle, *Quiscalus mexicanus*. The order *Passeriformes* contains a variety of birds, including larks, cardinals, crows and jays.

Palm Cockatoo

• ORDER •	• FAMILY •	• GENUS & SPECIES •
Psittaciformes	Cacatuidae	Probosciger aterrimus

KEY FEATURES

- Largest of all cockatoos — measures up to 27" in length

- Deftly cracks open nuts and the hard fruit of its favorite food, *Pandanus* palm, with its uniquely-shaped bill

- Cannot hiss like other cockatoos, but is adept at mimicking human language

- Captive birds live more than 30 years

WHERE IN THE WORLD?

Found in the tropical rainforest of New Guinea and in north Australia on the Cape York Peninsula; also on Misool, Yapen, West Papuan and Aru islands

LIFECYCLE

Unlike other cockatoos, the palm cockatoo does not congregate in large flocks; it roosts alone, feeds alone or in pairs and rarely forms groups of more than seven.

HABITAT

In Australia, the palm cockatoo makes its home in the eucalyptus forests of the Cape York Peninsula and the rainforests of the Aru Islands; in New Guinea, it populates the dense savannah woodlands in the northern and western sections of the island. Palm cockatoos inhabit wetter, warmer climates than most other cockatoo species and usually perch in leafless branches. Mating pairs maintain territories with several good nesting trees, which they check on periodically during the year.

▶ **MAJESTIC PERCH**
The watchful cockatoo scans the humid rainforest that it calls home.

BREEDING

Young palm cockatoos demand constant and attentive care, so adults form strong pair bonds and share parenting duties. Since they pair for life and remain close together throughout the year, courtship rituals are kept to a minimum, with only courtship-feeding the rule. Together, the male and female regularly inspect their territory for potential nesting sites. The ideal home is a hollow spot in a tree, 10–25' above the ground. Once they choose a site, they'll use it year after year, constructing a nest by shredding twigs and dropping them into the hollow (*left*).

Both parents incubate the single egg for about 30 days. Then the chick begins the difficult and exhausting hatching process, which can take 3–4 days; it finally emerges without down. The chick will stay in the nest for between 100–110 days, longer than any other parrot species. It will be two more weeks before the young cockatoo is ready to make its first flight.

CONSERVATION

The population of palm cockatoos is diminishing. In New Guinea the bird's rainforest habitat has been reduced due to logging; people also hunt the birds for food and capture them so they can be sold in other countries as pets. In Australia the palm cockatoo is protected and listed in Appendix I of CITES, yet many of these popular birds are smuggled out of the country each year.

LOG CABIN

❶ A suitable site...
Pairs mate for life; their first task as a mating pair is to find a safe, hollow tree in which to build their nest.

❷ Making the bed...
The two birds work together to build a nest by gathering sticks and then shredding and dropping them inside to form a platform.

❸ Baby makes three...
After a 30-day incubation, the female usually lays a single egg. Chicks remain in the safety of the nest for up to 110 days.

❹ Dual parenting
A young cockatoo is demanding: both parents work to feed and care for their offspring until six weeks after it leaves the nest.

FOOD & FEEDING

The humid rainforests offers the cockatoo an abundance of nuts, fruits, seeds and grubs. The bird shells and eats nuts efficiently: first, it husks the nut, turning it with its tongue while working the shell off with its beak. Then it splits the nut in two, storing the halves in its lower mandible. With a cache stored, the bird can push pieces forward with its tongue, split them on the edge of its lower beak and throw them back, with a quick toss of its head, to be swallowed.

▶ **SKILLED CRAFTSMAN**
The cockatoo sharpens its bill before feeding.

 **DID YOU KNOW?**

● Males sometimes exhibit a curious courtship behavior: they hold small sticks in their claws and drum on a hollow log.

● Cockatoos appear to be playful birds. They have been observed flying at one another, trying to knock each other from their perches.

 ## BEHAVIOR

Palm cockatoos roost separately, but call to each other after sunrise. After a brief foraging trip, a relatively small flock gathers in neutral territory. There they preen, display and interact socially. Long flights are usually necessary between roosting and feeding sites; the group disperses and often meets again at a convenient tree. At sunset, pairs leave the group and return to their own territories. After making a round of the nesting area, the two birds each roost separately. If predators or other birds intrude, a palm cockatoo stamps its feet and its facial patches blush a deep crimson.

◀ **PALM PEDICURE**
A cockatoo uses its bill to clean its foot and claws.

▲ **LOW BOW**
A cockatoo bends over during a social encounter.

PROFILE PALM COCKATOO

With its broad tail and impressive, fanlike head crest, the palm cockatoo appears quite large, but it's in fact relatively lightweight.

FEMALE
Males and females differ very little in appearance. The female has a slightly smaller upper mandible, only 2.9" as compared to 3.7" in males.

FACIAL PATCH
The patch of skin on the palm cockatoo's cheek ranges from beige to orange-pink to bright red, depending on the bird's environment and health; it flushes red when the cockatoo is stressed.

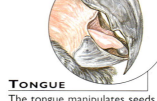

TONGUE
The tongue manipulates seeds during the husking process. The cockatoo can't close its beak completely, so it always shows its black-tipped, red tongue.

BILL
The upper mandible is larger than the lower and has several notches for holding and grinding different-sized nuts and fruits. The lower mandible has a cavity for storing halves of split seeds.

PLUMAGE
The palm cockatoo is the only cockatoo with all dark plumage, described by many as black but often appearing a slate gray.

FEET
The *zygodactyl* arrangement, with two toes pointing forward and two behind, allows a firm grip on branches as the cockatoo moves through the trees.

VITAL STATISTICS

WEIGHT	Male 1–2.5 lbs.; female 1–2 lbs.
LENGTH	19–27"
SEXUAL MATURITY	Unknown
BREEDING SEASON	October–December
NUMBER OF EGGS	1–2
INCUBATION PERIOD	30 days
FLEDGING PERIOD	Up to 135 days
BREEDING INTERVAL	1 year
TYPICAL DIET	Fruits, berries, seeds and grubs
LIFESPAN	Over 30 years, in captivity; unknown in the wild

RELATED SPECIES

● The palm cockatoo, often called the great black cockatoo, is the only species in its genus, *Probosciger*, but is 1 of 6 species of black cockatoo in 2 genera, including the glossy black cockatoo, *Calyptorhynchus lathami*. These birds join 14 species of white/gray cockatoo and 1 species of cockatiel in the family *Cacatuidae*. The cockatiel is the smallest in the family, weighing only 3 oz.

 CREATURE COMPARISONS

Unlike the palm cockatoo, the red-tailed black cockatoo (*Calyptorhynchus banksii*) has adapted to the drier conditions in the woodlands and shrublands of northern Australia and the open forests of the southeastern part of the continent. Measuring about 25" in length, the red-tailed black cockatoo is similar in size to the palm cockatoo. It lacks the palm cockatoo's head crest, but the male red-tailed cockatoo distinguishes itself with splashy red panels in its tail, hence its name; the female has orange and yellow spots and bars.

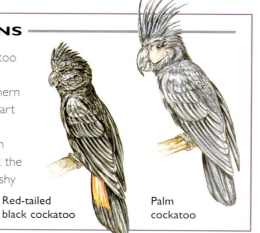

Red-tailed black cockatoo

Palm cockatoo

Paradise Whydah

• ORDER •	• FAMILY •	• GENUS & SPECIES •
Passeriformes	*Ploceidae*	*Vidua paradisaea*

KEY FEATURES

- Known as a brood parasite: lays its eggs in the nests of the green-winged pytilia
- Whydah chicks mimic the calls and actions of their foster siblings
- The male's elongated tail feathers make him more than twice the length of the female

WHERE IN THE WORLD?

Found over much of eastern and southern Africa, from Angola east through Zambia and Zimbabwe east to Mozambique

LIFECYCLE

The paradise whydah not only relies on the green-winged pytilia for hatching and feeding its chicks, but males mimic the pytilia's song when courting female whydahs.

HABITAT

The paradise whydah frequents open savannahs in eastern and southern Africa with scattered acacia trees or shrubs. The bird can be found at elevations from sea level to 7,000', but the largest concentrations are at altitudes below 5,000'. Since the whydah is dependent on its host, the green-winged pytilia, during breeding, the whydah seeks the same type of dry, open country woodlands and savannahs with thorny scrub that the pytilia prefers.

▶ **TREE HOUSE**
The paradise whydah prefers the open woodlands and savannahs of eastern and southern Africa.

FOOD & FEEDING

The whydah wanders along the open savannah in search of insects and seeds, which the bird cracks open with its strong bill. While walking, the whydah often jumps backward with both feet together, scratching the ground to uncover a tasty snack. The bird feeds in small or mixed flocks of whydahs.

▶ **SCRATCH AND SNIFF**
Scratching with its feet, the whydah unearths seeds.

BREEDING

During the breeding season, the polygamous male flies with his impressively long tail feathers raised almost at a right angle to his body. The male hovers around a female, slowly beating his wings as if to hypnotize his potential mate. All whydahs are brood parasites and lay their eggs in the nests of other birds. The paradise whydah males shows the female the nest of a pytilia, and the female then lays 1–5 eggs in the pytilia's nest. Once the eggs hatch, after incubating for 12-13 days, the whydah's chicks mimic the calls of their fellow nestlings. The pytilias then take over the care of the young whydahs until they fledge about 16 days later.

? DID YOU KNOW?

- Adult paradise whydahs are 60% heavier than the green-winged pytilia.

- Paradise whydah chicks are actually more grayish and dark-skinned, with broader bills than their pytilia "siblings."

- Whydah eggs are only 0.1" larger than the green-winged pytilia's eggs.

- The paradise whydah parasitizes 92% of the nests of its host; there is a maximum combined clutch of 10 eggs.

702 Paradise Whydah

PARASITIC NESTER

① Unknowing host...
The green-winged pytilia, which builds its grass nest in a thorn bush, will soon return to an unexpected surprise.

② A home of their own...
The female paradise whydah closely inspects the nest, which already contains one white pytilia egg, before laying her own eggs.

③ Then there were four...
The pytilia doesn't seem to mind the new additions, since they closely resemble her eggs in size, shape and color.

④ Mimic this, mimic that
Not even the parents can tell the difference between the chicks since the markings in the mouth openings are so similar.

BEHAVIOR

The mimicking behavior of the whydah begins at birth, when hungry whydah chicks mimic the vocalizations of their foster siblings. Adult males also mimic the song, chirps and warbling sounds of the green-winged pytilia, especially during the breeding season. The whydah has its own calls as well that begin with a long, introductory whistle. The male mixes its own song with the mimicked song during the mating ritual. During courtship, the territorial male will perform spectacular cruising flights high in the air, which end in a series of swooping plunges back to treetops where the females gather. Despite its small size, the whydah is a pugnacious bird that will chase other larger birds away if threatened. Small flocks that forage and perch together will eventually separate into pairs during the breeding season.

▶ **COPY CAT**
The male paradise whydah mimics the green-winged pytilia with its calls.

CONSERVATION

The paradise whydah is fairly common and can be found in large numbers throughout its African range. A high rate of hatching success assures the whydah continued strong populations. As with any brood parasite, the success of its host, the green-winged pytilia, is essential to the paradise whydah's success. The green-winged pytilia is also fairly plentiful and not threatened at this time.

PROFILE PARADISE WHYDAH

During the breeding season, the male paradise whydah is unmistakable with his black plumage and spectacular sweeping tail.

BILL
The large, dark bill is short and conical, well adapted for cracking seeds.

FEMALE
The female's mottled brown and buff plumage is similar to but duller than that of the nonbreeding male. Her breast is either plain or necklaced with fine streaks; she has C-shaped marks on the side of the face.

NON-BREEDING PLUMAGE
During the non-breeding seasons — June to November in South Africa and April to October in Kenya — the male has a black and cream-streaked face and head. Its back has shades of tan and black. The male will also molt and lose its long tail feathers.

BREEDING PLUMAGE
The male's tail feathers grow extraordinarily long — up to 13" — during the breeding season; his plumage becomes distinct from the female's, but not until the age of 2. The body is black, the breast is chestnut and the broad buff nape band is golden.

FEET
Three toes point forward and the hind toe (*hallux*) points backward; this helps the bird secure its foothold on a tree branch or other perch.

VITAL STATISTICS

WEIGHT	Up to 0.75 oz.
LENGTH Head/Body Tail	About 4–5" Up to 13" in male; about 1" in female
SEXUAL MATURITY	1–2 years
BREEDING SEASON	Varies with location
NUMBER OF EGGS	1–5
INCUBATION PERIOD	12–13 days
FLEDGING PERIOD	About 21 days
BREEDING INTERVAL	Up to 5 clutches per breeding season
TYPICAL DIET	Seeds and insects
LIFESPAN	Unknown

RELATED SPECIES

● The paradise whydah is one of nine species in the genus *Vidua*. The genus includes the straw-tailed whydah, *Vidua fischeri*, and the broad-tailed paradise whydah, *V. obtusa*. Within the family Ploceidae, there are 143 species of weaver, sparrow and snow finch in 18 genera. The family contains the village weaver, *Ploceus cucullatus*, and the house sparrow, *Passer domesticus*.

CREATURE COMPARISONS

The male pin-tailed whydah (*Vidua macroura*) measures up to 12" in length, a bit smaller than the paradise whydah. The breeding male is black and white and has four long, central tail feathers. He is called "King-of-Six" for his ability to court six females at once. The pin-tailed whydah inhabits cultivated areas and gardens in most parts of southern Africa, Kenya and northern Tanzania. It relies on two host species for nesting, the common black-rumped waxbill and orange-tailed waxbill, unlike the paradise whydah, which lays eggs in the nests of only one host species.

Pin-tailed whydah

Paradise whydah

Pel's Fishing Owl

• ORDER •	• FAMILY •	• GENUS & SPECIES •
Strigiformes	Strigidae	*Scotopelia peli*

KEY FEATURES

● Does not rely on sound to find prey, as most owls do

● Loose feathering on head gives it a distinctive fluffy appearance

● Unfeathered feet are equipped with sharp scales and talons to catch and hold fish

● Lacks the ear tufts found in most owl species

WHERE IN THE WORLD?

Found along large rivers and tree-lined streams of Africa, from Nigeria across the Congo Basin and central Africa to the Zambezi River and in Ethiopia south to South Africa

LIFECYCLE

The large, orange-colored Pel's fishing owl becomes especially noisy before breeding; mating pairs engage in distinctive duets that last for several minutes.

HABITAT

▲ TREE HOUSE
The fishing owl sleeps, roosts, nests and hunts in the serenity of the trees throughout its range.

Living along the edges of rivers and wetlands, the Pel's fishing owl resides throughout most of tropical Africa south of the Sahara Desert. It can be found in strips of riverine forests as well as the great tropical forests of the Congo River.

This owl thrives amid large, ancient trees, which provide numerous cavities for roosting and nesting as well as long branches extending over the water — a perfect fishing perch. Rarely seen, Pel's owl is secretive and can go undetected for long periods of time because it only comes out at night. The Pel's fishing owl spends the majority of its time in trees and rarely descends to the ground, which is why little is known about this mysterious owl's way of life.

? DID YOU KNOW?

- First captured in 1850 in Ghana (then called the Gold Coast), the Pel's fishing owl was named for that country's governor, H.S. Pel.
- Murky waters force the owl to listen, not watch, for fish movement.
- Starvation accounts for most of the deaths of the Pel's fishing owl, particularly among inexperienced juvenile birds, which can't hunt well.
- The Pel's owl only eats live prey, and always consumes it head first; dead fish are ignored.

FOOD & FEEDING

The Pel's fishing owl doesn't rely on the same hunting methods that most owls use. For example, in most owls facial disks around each eye funnel even the slightest sound into the ear openings, which are located just behind the eye. With less prominent disks, the Pel's owl relies more on sight than sound when hunting for its prey. Also, most owls are stealthy, silent fliers. The Pel's owl, with its specialized diet of fish, does not need noiseless flight, since fish cannot hear the owl approaching.

The Pel's owl perches on large branches that overhang the water, ready to attack. Once a fish is spotted, the owl swoops down, feet first, and snatches it from the surface of the water with its talons, never plunging into the water. Common targets include bream, catfish, and barbel. The average catch weighs less than 1 lb.

▶ MIGHTY FISHERMAN
This owl is perfectly equipped for landing fish.

BREEDING

The Pel's fishing owl heralds the breeding season with distinctive melodic duets designed to attract a mate. Once paired, mating begins in the summer to coincide with peak heights of nearby rivers. Chicks are reared as the water levels recede, when fishing conditions are most favorable. The Pel's fishing owl does not build a nest; the female usually lays her eggs in the hollow of a large tree.

The female incubates the eggs alone for about 35 days, but the male stays close and feeds her as well as the young once they hatch. Although the Pel's fishing owl lays two eggs, only one chick is reared. The first chick hatches up to five days before the second and bullies its younger sibling until it dies of starvation. The surviving chick fledges in about 70 days, but may stay with its parents longer.

During long periods of drought, a mated pair will sometimes let a year or more pass before breeding.

▶ **THE APPRENTICE**
The young owl learns to hunt from its parents.

CONSERVATION

The Pel's fishing owl is rarely seen, and the size of its population is unknown, but experts do not yet consider it to be endangered. It is listed in Appendix II of CITES (Convention on International Trade in Endangered Species). However, the populations of rufous and vermiculated fishing owls, the other two African species, are endangered. Habitat destruction is the chief cause.

BEHAVIOR

Being mainly nocturnal, Pel's fishing owls roost alone or in pairs during the day and emerge around dusk to eat. They have such keen eyesight that they can see the ripples in the water made by fish. The Pel's owls are extremely territorial, utilizing distinct calls to warn off intruders and defend their turf.

DRIVEN TO DISTRACTION

❶ Feeding...
A pair of Pel's fishing owls nests in an abandoned eagle's nest. The female incubates the eggs while the male hunts for their food.

❷ Chasing...
Vigorously defending his family, the male owl chases away an African fish eagle that has come to raid the nest.

❸ Guarding...
Standing guard near the nest, the male fish owl catches a glimpse of a blotched genet venturing a little too close.

❹ Distracting
The owl dives off the branch into the undergrowth, thrashing about as if injured. Falling for the display, the genet is lured from the nest.

PROFILE   Pel's Fishing Owl

One of the largest owls in the world, the uncommonly strong Pel's fishing owl can subdue a wiggling fish virtually equal to its own weight.

PRIMARY FEATHERS
Stiff wing feathers (A) lack the soft fringe on their leading edges (B) that enable most owls, such as the barn owl, to fly silently.

BILL
The strongly curved bill is large, with a broad base and, a sharp tip that is used for ripping and tearing prey.

PELLET
The owl has no crop for storage; food is swallowed whole and enters the stomach. All indigestible matter, such as feathers and bone, is then regurgitated as large pellets.

FEET
The foot is unfeathered with long, spiny toes and highly arched talons to grip slippery fish.

VITAL STATISTICS

WEIGHT	4–5 lbs.
LENGTH	20–24"
WINGSPAN	4.5–5'
SEXUAL MATURITY	Unknown
BREEDING SEASON	February–April
NUMBER OF EGGS	2, but only 1 hatchling is reared
INCUBATION PERIOD	About 35 days
FLEDGING PERIOD	About 70 days
BREEDING INTERVAL	1 year
TYPICAL DIET	Mainly fish; also crabs, frogs and small mammals
LIFESPAN	Unknown

RELATED SPECIES

● The fishing owls are confined to Africa and Asia, and comprise 7 species in 2 genera. The genus *Scotopelia* contains all 3 African species: the Pel's fishing owl; the rufous fishing owl, *S. ussheri*; and the vermiculated fishing owl, *S. bouvieri*. The family Strigidae contains about 175 species of raptors that are found in virtually every place in the world.

CREATURE COMPARISONS

Most fishing owls have unfeathered legs and feet, but the Blakiston's fish owl *(Ketupa blakistoni)* is fully feathered to cope with the cold northerly climates of China and Russia. Inhabiting cense forests bordered by rivers, lakes or the sea, Blakiston's owl seeks out fast-flowing water that does not freeze in winter. Its plumage is dark brown with wavy horizontal barring, compared to the rufous-orange color of the Pel's fishing owl.

Pel's fishing owl   Blakiston's fish owl

Peregrine Falcon

· ORDER ·
Falconiformes

· FAMILY ·
Falconidae

· GENUS & SPECIES ·
Falco peregrinus

KEY FEATURES

● Hunts other birds in the air; possesses the agility and flying skills to catch even the most aerobatic species

● Specializes in spectacular diving attacks at speeds of more than 100 mph

● Kills victims by the devastating impact of its strike or dispatches its quarry on the ground with a bite to the back of the neck

WHERE IN THE WORLD?

Breeds in northern North America and Eurasia; overwinters farther south in the Americas, Africa south of the Sahara, parts of Europe, Southeast Asia and Australia; some populations reside year-round

LIFECYCLE

The peregrine falcon is renowned for its breathtaking dives on other birds, plummeting from high altitudes to burst without warning upon its hapless victims.

HABITAT

▲ **Vantage point**
High up on its cliffside aerie (nest site), a falcon surveys its domain.

▼ **Far and away**
The widespread peregrine falcon adapts to a variety of habitats, but favors remote, open regions with high cliffs.

The peregrine falcon is the most widespread of all falcons, occurring on all continents except Antarctica.

The falcon lives in a wide variety of habitats, but avoids wetlands, such as marshes. It prefers open land bordered by high cliffs, especially along rocky coasts. It also breeds inland in gorges and disused quarries.

The bird can also be seen in large U.S. cities, where it sometimes nests on top of skyscrapers and preys on pigeons. The peregrine nests in trees in some of the Baltic countries and Australia and in church steeples or other tall buildings in parts of Europe.

BEHAVIOR

The peregrine falcon is one of the world's fastest birds. On its dives, it angles its wings backward and pulls them close to its body to form an efficient aerodynamic profile. It may free-fall 1,000' or more, slicing through the sky at speeds near 100 mph. In level flight, the peregrine flies much more slowly and may even be outpaced by some species of duck.

▲ **Tailing prey**
The peregrine reaches its highest velocities in dives, but it also exhibits speed and power in flat pursuit.

Peregrines from tropical, subtropical and some island locations tend to be year-round residents, but individuals from the most northerly parts of their range migrate in the autumn to avoid the snows of winter. Those breeding in northern Canada, for example, may fly 4,800 miles to southern South America and back again each year — a round trip greater than that of any other falcon species.

DID YOU KNOW?

● The name "peregrine" stems from a Latin word meaning "foreigner or wanderer." A peregrine on migration was recorded to cover 960 miles in 48 hours.

● The peregrine's eyes are larger and heavier than a human's. It can spot small birds on the ground while flying at heights of more than 1,000'.

CONSERVATION

Peregrine populations declined from the 1950s to the 1970s as a result of the pesticide DDT. Many falcons died or laid eggs with thin shells after preying on seed-eating birds with deadly levels of DDT. Today, DDT is banned in many countries, and some populations are on the rise.

FOOD & HUNTING

The peregrine preys almost exclusively on birds and often catches them on the wing. Its diet depends on what's available within its range. Falcons in coastal locations, for example, prey mainly on seabirds around their cliffside nesting colonies.

The peregrine can use several hunting techniques. This hunter sometimes snatches a resting bird from an exposed perch in the treetops or seizes a victim on the wing using its incredible speed to outfly slower and less agile birds. Of all the techniques, however, the diving attack is the most dramatic.

The falcon typically flies at a great height while hunting. After spotting potential prey, it launches into a dive or "stoop." Once committed to its course, the falcon can't change direction, and a high proportion of dives fail as a result. But when the falcon gets it right, the strike is devastating. The victim is dealt a terrible, often lethal, blow without ever sensing its impending doom.

BOLT FROM THE BLUE

❶ Patrol...
Patrolling high in the skies, the peregrine falcon glides and soars, searching for signs of low-flying prey far below.

❷ Dive...
The falcon targets a duck. Angling its tapered wings sharply back, it plunges into a free-falling dive, reaching a speed of 100 mph.

❸ Strike...
The falcon thrusts its sharp talons forward just before impact. The prey is struck with such devastating force that it is killed instantly.

❹ Retrieve
The falcon brakes rapidly, turns and swoops again, snatching its victim up before its broken body tumbles to the ground.

BREEDING

Open sky is the stage for falcon courtship. Here, the bird plays out the full repertoire of its aerobatics in a showy display of looping, swooping and figure-eight flights as a prelude to mating. The male often passes prey to his mate in flight as a courtship offering.

The peregrine doesn't build a nest itself, but uses the old nest of another bird. The female lays three or four eggs and performs most of the incubation. When chicks hatch, the male continues hunting to provide food. At about four weeks, the chicks begin to show juvenal plumage and may take their first flight two weeks later. Parents continue to feed juveniles for up to two months, even after they've started to fly, often passing food to them on the wing.

Young sharpen flying skills by play-fighting in the air or by "dive-bombing" other birds. Eventually they leave the nest, driven away by their parents or in response to an instinctive urge to move on.

Profile — Peregrine Falcon

The peregrine is a formidable predator of the skies, superbly equipped to seek and intercept other birds in flight, like a living air-to-air missile.

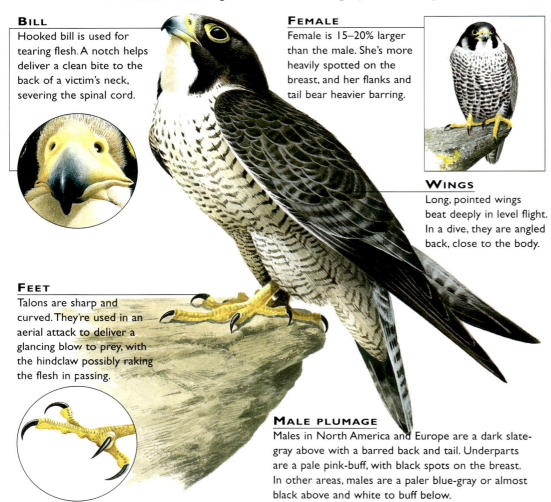

BILL
Hooked bill is used for tearing flesh. A notch helps deliver a clean bite to the back of a victim's neck, severing the spinal cord.

FEMALE
Female is 15–20% larger than the male. She's more heavily spotted on the breast, and her flanks and tail bear heavier barring.

WINGS
Long, pointed wings beat deeply in level flight. In a dive, they are angled back, close to the body.

FEET
Talons are sharp and curved. They're used in an aerial attack to deliver a glancing blow to prey, with the hindclaw possibly raking the flesh in passing.

MALE PLUMAGE
Males in North America and Europe are a dark slate-gray above with a barred back and tail. Underparts are a pale pink-buff, with black spots on the breast. In other areas, males are a paler blue-gray or almost black above and white to buff below.

CREATURE COMPARISONS

The peregrine is one of many swift and deadly birds of prey in the family *Falconidae*. The merlin (a small falcon) often charges its prey in a low-level attack. The common kestrel, smaller than the peregrine, is the most numerous falcon. The gyrfalcon, largest of all falcons, is more heavily built than other species and has broader wings.

Merlin • Common kestrel • Peregrine falcon • Gyrfalcon

VITAL STATISTICS

Weight	19–53 oz.; female larger than male
Length	17–20"
Wingspan	32–48"
Sexual Maturity	Usually 2–3 years
Breeding Season	February–August in northern parts of range
Number of Eggs	3–4
Incubation Period	29–32 days
Fledging Period	About 6 weeks
Breeding Interval	1 year
Typical Diet	Birds up to the size of a duck; small mammals
Lifespan	About 18 years

RELATED SPECIES

● The family *Falconidae* contains about 40 falcon species, including the peregrine falcon's close relative, Barbary falcon (or black shaheen), *Falco pelegrinoides*, and the endangered Mauritius kestrel, *Falco punctatus* (below).

DISCARD

Genesee District Libraries
Fenton - Jack R. Winegarden Children's Center
200 E. Caroline St.
Fenton, MI 48430

598.03 E

The Encyclopedia of birds

9-07